Mechanical Engineering Series

Frederick F. Ling
Series Editor

George Chryssolouris

Laser Machining
Theory and Practice

With 164 Illustrations

Springer Science+Business Media, LLC

George Chryssolouris
Laboratory for Manufacturing and Productivity
Room 35-134
Massachusetts Institute of Technology
Cambridge, Massachusetts 02139

ISBN 978-1-4757-4086-8   ISBN 978-1-4757-4084-4 (eBook)
DOI 10.1007/978-1-4757-4084-4

Library of Congress Cataloging-in-Publication Data
Chryssolouris, G. (George)
    Laser machining : theory and practice / George Chryssolouris.
      p.   cm. — (Mechanical engineering series)
    Includes bibliographical references.

    1. Laser beam cutting. I. Title. II. Series: Mechanical
engineering series (Berlin, Germany)
TJ1191.C46   1991
671.3'5—dc20   90-26508

Printed on acid-free paper

Series Preface

Mechanical engineering, an engineering discipline born of the needs of the industrial revolution, is once again asked to do its substantial share in the call for industrial renewal. The general call is urgent as we face profound issues of productivity and competitiveness that require engineering solutions, among others. The Mechanical Engineering Series is a new series, featuring graduate texts and research monographs, intended to address the need for information in contemporary areas of mechanical engineering.

The series is conceived as a comprehensive one that will cover a broad range of concentrations important to mechanical engineering graduate education and research. We are fortunate to have a distinguished roster of consulting editors, each an expert in one of the areas of concentration. The names of the consulting editors are listed on the first page of the volume. The areas of concentration are applied mechanics, biomechanics, computational mechanics, dynamic systems and control, energetics, mechanics of materials processing, thermal science, and tribology.

Professor Wang, the consulting editor for processing, and I are pleased to present the sixth volume of the series: *Laser Machining: Theory and Practice* by Professor Chryssolouris. This is the first graduate textbook to address this area which is at the forefront of manufacturing technology. The selection of this volume underscores again the interest of the Mechanical Engineering Series to provide our readers with topical graduate texts as well as research monographs.

New York, New York   *Frederick F. Ling*

Preface

This book attempts to address laser machining in a way that is useful for researchers and academicians alike who are interested in state-of-the-art in the field. It also attempts to help practitioners, particularly manufacturing engineers, who are considering laser machining as a solution to the machining needs of their factories and plants.

To put laser machining in its correct context among manufacturing processes, the *first chapter* provides an overview of conventional material removal processes. The *second chapter* deals with the lasers themselves. *Chapter 3* covers laser machining systems. *Chapter 4* provides an overview of necessary knowledge from heat transfer and fluid mechanics, which is required in order to understand the physical mechanisms and the analysis of a thermal process such as laser machining. *Chapter 5* includes analyses of one, two and three-dimensional laser machining processes based on the fundamentals of heat transfer and fluid mechanics discussed in Chapter 4. *Chapter 6*, the final chapter, summarizes state-of-the-art laser machining applications in research laboratories and industrial practice.

I would like to thank Professor David Wormley, the head of the Mechanical Engineering Department, for his support and encouragement. I am also very indebted to the National Science Foundation, whose funding made this work possible, particularly to Drs. Branimir Von Turkovich, John Meyer, Marvin DeVries, Ranga Komanduri, and Michael Wozny. The work was also sponsored by a number of industrial firms. Among them, I would like to thank Ford Motor Company, Coherent General, the Cross Company, and Gleason Works. Professor Nam Suh, the founder of M.I.T.'s Laboratory for Manufacturing and Productivity, although not directly involved in laser machining, has provided me with continued intellectual guidance and inspiration over the years. Numerous graduate and undergraduate students have also worked with me on a variety of laser machining projects. Some come from the Laser Center of the University of Hannover in Germany, directed by Professor Hans Kurt Tönshoff, to whom I would also like to express my thanks.

Paul Sheng, a graduate student in the Department of Mechanical Engineering at M.I.T. who has been working with me towards his Ph.D. in laser machining, was instrumental in putting this book together. He contributed substantially to the text, the figures, the experimental results, and the overall layout of the book. His excellent contributions and his friendly and helpful attitude toward his fellow researchers and myself were great assets during the creation of this manuscript.

I would also like to thank the many students who were involved in preparing, criticizing and correcting this manuscript, among them Woo Chun Choi, Stanley Kyi, James Bredt, Edward Wilson, Joe Kwasnoski, and Mike Domroese. This book would not have been possible without the substantial help of Ms. Jennifer Gilden, who prepared most of the figures, edited and proofread the manuscript numerous times, and tirelessly worked on the final version of the text. I certainly appreciate her effort and express my sincere thanks. Finally, I would like to thank my family for their understanding of my *long* hours in preparing such a *short* monograph.

George Chryssolouris

September 1990
Cambridge, Massachusetts

Acknowledgments

I would like to acknowledge the following publishers and authors for the use of a number of different figures and tables:

Fig. 1.9	Reprinted from <u>Machining Datability Handbook</u>, 3rd edition, by permission of the Machinability Data Center. © 1980 by Metcut Research Associates Inc.
Tables 1.2 - 1.4	Reprinted from <u>Machining Datability Handbook</u>, 3rd edition, by permission of the Machinability Data Center. © 1980 by Metcut Research Associates Inc.
Fig. 1.10	Reprinted from <u>Machining Datability Handbook</u>, 3rd edition, by permission of the Machinability Data Center. © 1980 by Metcut Research Associates Inc.
Fig. 2.1	from <u>Lasers: Operation, Equipment, Application and Design</u>, by the staff of Coherent, Inc. (New York: McGraw-Hill, 1980). Reprinted courtesy of Coherent, Inc.
Fig. 2.2	from "Advances in Optical Masers," by A. Schawlow. Copyright © July 1963 by Scientific American, Inc. All rights reserved.
Fig. 2.3	from "Laser Light," by A. Schawlow. Copyright © Sept. 1968 by Scientific American, Inc. All rights reserved.
Fig. 2.8	from <u>Lasers: Operation, Equipment, Application and Design</u>, by the staff of Coherent, Inc. (New York: McGraw-Hill, 1980). Reprinted courtesy of Coherent, Inc.
Fig. 2.9	from <u>Lasers: Operation, Equipment, Application and Design</u>, by the staff of Coherent, Inc. (New York: McGraw-Hill, 1980). Reprinted courtesy of Coherent, Inc.
Fig. 2.10	from "High Power Carbon Dioxide Lasers," by C.K.N. Patel. Copyright © August 1968 by Scientific American, Inc. All rights reserved.
Fig. 2.11	from <u>Lasers: Operation, Equipment, Application and Design</u>, by the staff of Coherent, Inc. (New York: McGraw-Hill, 1980). Reprinted courtesy of Coherent, Inc.

Fig. 2.12   from Lasers: Operation, Equipment, Application and Design, by the staff of Coherent, Inc. (New York: McGraw-Hill, 1980). Reprinted courtesy of Coherent, Inc.

Fig. 2.18   from Lasers: Operation, Equipment, Application and Design, by the staff of Coherent, Inc. (New York: McGraw-Hill, 1980). Reprinted courtesy of Coherent, Inc.

Fig. 3.13   from Lasers: Operation, Equipment, Application and Design, by the staff of Coherent, Inc. (New York: McGraw-Hill, 1980). Reprinted courtesy of Coherent, Inc.

Fig. 3.14   from Lasers: Operation, Equipment, Application and Design, by the staff of Coherent, Inc. (New York: McGraw-Hill, 1980). Reprinted courtesy of Coherent, Inc.

Figs. 3.15 -3.17   from Laser Lab product literature. Reprinted courtesy of Laser Lab.

Fig. 3.21   from "Circular and Non-Circular Nozzle Exits from Supersonic Gas Jet Assist in CO_2 Laser Cutting," by J. Fieret and B.A. Ward. Proceedings, *Third International Conference on Lasers in Manufacturing*, 1986. Reprinted courtesy of Springer-Verlag, Heidelberg.

Fig. 3.24   from "A Smart Laser Cutter," by B. Burg et al. Proceedings, High Power Lasers and Their Industrial Applications, Vol. 650, 1986. Reprinted courtesy of SPIE.

Fig. 4.15   from Theory and Problems of Heat Transfer, by D.R. Pitts and L.E. Sissom (New York: McGraw-Hill, Inc., 1977). Reproduced with permission of McGraw-Hill, Inc.

Fig. 4.16   from Theory and Problems of Heat Transfer, by D.R. Pitts and L.E. Sissom (New York: McGraw-Hill, Inc., 1977). Reproduced with permission of McGraw-Hill, Inc.

Fig. 4.18   from Theory and Problems of Heat Transfer, by D.R. Pitts and L.E. Sissom (New York: McGraw-Hill, Inc., 1977). Reproduced with permission of McGraw-Hill, Inc.

Fig. 4.22   from Theory and Problems of Heat Transfer, by D.R. Pitts and L.E. Sissom (New York: McGraw-Hill, Inc., 1977). Reproduced with permission of McGraw-Hill, Inc.

Fig. 6.2   from "Study of Affecting Parameters in Laser Hole Drilling of Sheet Metals," by B.S. Yilbas. *Journal of Engineering Materials and Technology* (ASME), Vol. 109, Oct. 1987. Reprinted courtesy of the ASME.

Fig. 6.3   from <u>Lasers: Operation, Equipment, Application and Design</u>, by the staff of Coherent, Inc. (New York: McGraw-Hill, 1980). Reprinted courtesy of Coherent, Inc.

Fig. 6.4   from "Werkstoffbearbeitung mit Laserstrahlung," by D. Petring et al. *Feinwerktechnik*, Vol. 96, 1988. Reprinted courtesy of Carl Hanser Verlag, Munich.

Fig. 6.5   from "Laser Cutting," by D. Schuocker. <u>Industrial Laser Annual Handbook</u>, 1986. Reprinted courtesy of PennWell Publishing Co.

Fig. 6.6   from "Gas-Jet Laser Cutting (Review)," by V. Babenko et al. *Soviet Journal of Quantum Mechanics*, Vol. 2, No. 5, March-April 1973. Reprinted courtesy of the American Institute of Physics.

Fig. 6.7   from "Laser Beam Cutting of Thick Steel," by G. Sepold and R. Rothe. *ICALEO*, 1983.

Fig. 6.8   from "Laser Beam Cutting of Thick Steel," by G. Sepold and R. Rothe. *ICALEO*, 1983.

Fig. 6.9   from "Laser Cutting of Thin Materials," by S. Engel. <u>Lasers in Modern Industry</u>, 1979. Reprinted courtesy of the Society of Manufacturing Engineers. Copyright 1974.

Fig. 6.10   from "Laser Cutting of Thin Materials," by S. Engel. <u>Lasers in Modern Industry</u>, 1979. Reprinted courtesy of the Society of Manufacturing Engineers. Copyright 1974.

Fig. 6.11   from "CO_2 Laser Marking," by D.C. Hamilton. <u>Industrial Laser Annual Handbook</u>, 1986. Reprinted courtesy of PennWell Publishing Co.

Fig. 6.12   from "YAG Laser Marking," by G. Garman and J. Ponce. <u>Industrial Laser Annual Handbook</u>, 1986. Reprinted courtesy of PennWell Publishing Co.

Fig. 6.13   from "YAG Laser Marking," by G. Garman and J. Ponce. <u>Industrial Laser Annual Handbook</u>, 1986. Reprinted courtesy of PennWell Publishing Co.

Fig. 6.17   from "Laser Machining of Ceramic," by A. Laudel. Department of
                   Energy Report No. BDX-613-2507, 1980. This work was prepared
                   for the United States Department of Energy under contract number
                   DE-AC044-76-DP-00613. Reprinted courtesy of Allied Signal Inc.,
                   Kansas City Division.

Fig. 6.18   from "Laser Machining of Ceramic," by A. Laudel. Department of
                   Energy Report No. BDX-613-2507, 1980. This work was prepared
                   for the United States Department of Energy under contract number
                   DE-AC044-76-DP-00613. Reprinted courtesy of Allied Signal Inc.,
                   Kansas City Division.

Fig. 6.19   from "Laser Machining of Ceramic and Silicon," by C. Hamann and
                   H. Rosen. High Power Lasers and Their Industrial Applications,
                   Vol. 801, 1987. Reprinted courtesy of SPIE.

Fig. 6.21   from "Machining of High Performance Ceramics and Thermal Etching
                   of Glass by Laser," by R. Harrysson and H. Herbertsson.
                   Proceedings, *Fourth International Conference on Lasers in
                   Manufacturing*, 1987. Reprinted courtesy of Elsevier Science
                   Publishers B.V./Physical Sciences & Engineering Division.

Fig. 6.28   from Lasers: Operation, Equipment, Application and Design, by the
                   staff of Coherent, Inc. (New York: McGraw-Hill, 1980). Reprinted
                   courtesy of Coherent, Inc.

Fig. 6.33   from "Machining of Fibre Reinforced Plastics," by W. Konig et al.
                   Annals of the CIRP, N2, 1985. Reprinted courtesy of CIRP.

Fig. 6.34   from "Characteristics of Laser Cutting Kevlar Laminates," by R.A.
                   Van Cleave. Department of Energy Report No. BDX-613-2075,
                   1979. This work was prepared for the United States Department of
                   Energy under contract number DE-AC044-76-DP-00613. Reprinted
                   courtesy of Allied Signal Inc., Kansas City Division.

Fig. 6.35   from "Characteristics of Laser Cutting Kevlar Laminates," by R.A.
                   Van Cleave. Department of Energy Report No. BDX-613-2075,
                   1979. This work was prepared for the United States Department of
                   Energy under contract number DE-AC044-76-DP-00613. Reprinted
                   courtesy of Allied Signal Inc., Kansas City Division.

Fig. 6.36   from "Machining of Fibre Reinforced Materials with Laser Beam: Cut
                   Quality Evaluation," by V. Tagliaferri et al. Proceedings, *Sixth
                   International Conference on Composite Materials*, 1987. Reprinted
                   courtesy of Elsevier Science Publishers B.V./Physical Sciences &
                   Engineering Division.

Fig. 6.37   from "Machining of Fibre Reinforced Plastics," by W. Konig et al.
                 Annals of the CIRP, N2, 1985. Reprinted courtesy of CIRP.

Fig. 6.38   from "Machining of Fibre Reinforced Materials with Laser Beam: Cut
                 Quality Evaluation," by V. Tagliaferri et al. Proceedings, *Sixth
                 International Conference on Composite Materials*, 1987. Reprinted
                 courtesy of Elsevier Science Publishers B.V./Physical Sciences &
                 Engineering Division.

Fig. 6.39   from "Cutting of Fibre-Reinforced Polymers with CW CO_2 Laser,"
                 by M. Flaum and T. Karlsson. High Power Lasers and Their
                 Industrial Applications, Vol. 801, 1987. Reprinted courtesy of SPIE.

Fig. 6.40   from "Investigation of Laser Grooving for Composite Materials," by
                 G. Chryssolouris, P. Sheng and W.C. Choi. Annals of the CIRP,
                 1988. Reprinted courtesy of CIRP.

Fig. 6.41   from "Investigation of Laser Grooving for Composite Materials," by
                 G. Chryssolouris, P. Sheng and W.C. Choi. Annals of the CIRP,
                 1988. Reprinted courtesy of CIRP.

Fig. 6.42   from "Characteristics of Laser Cutting Kevlar Laminates," by R.A.
                 Van Cleave. Department of Energy Report No. BDX-613-2075,
                 1979. This work was prepared for the United States Department of
                 Energy under contract number DE-AC044-76-DP-00613. Reprinted
                 courtesy of Allied Signal Inc., Kansas City Division.

Fig. 6.45   from Lasers: Operation, Equipment, Application and Design, by the
                 staff of Coherent, Inc. (New York: McGraw-Hill, 1980). Reprinted
                 courtesy of Coherent, Inc.

Fig. 6.46   from Lasers: Operation, Equipment, Application and Design, by the
                 staff of Coherent, Inc. (New York: McGraw-Hill, 1980). Reprinted
                 courtesy of Coherent, Inc.

Fig. 6.47   from Laser Processing and Analysis of Materials, by W.W. Duley,
                 1983. Reprinted courtesy of Plenum Press.

Fig. 6.48   from Lasers: Operation, Equipment, Application and Design, by the
                 staff of Coherent, Inc. (New York: McGraw-Hill, 1980). Reprinted
                 courtesy of Coherent, Inc.

Fig. 6.49   from Lasers: Operation, Equipment, Application and Design, by the
                 staff of Coherent, Inc. (New York: McGraw-Hill, 1980). Reprinted
                 courtesy of Coherent, Inc.

Figs. 6.50 - 6.52   from "An Introduction to Thick Film Resistor Trimming by Laser," by M. Oakes. <u>Lasers in Modern Industry</u>, 1979 (Society of Manufacturing Engineers). Reprinted courtesy of SPIE.

Fig. 6.53   from "Laser Production Applications in Microelectronics," by D.L. Parker. <u>Industrial Laser Annual Handbook</u>, 1986 (PennWell Publishing Co.) Reprinted by permission of Texas Instruments.

Contents

1
Overview of Machining Processes

This chapter summarizes some commonly used manufacturing processes so the potential user of the laser machining process can view it in its proper relation to the family of material removal processes. The chapter includes a review of common processes such as turning, drilling, milling and grinding, and provides a summary of attainable material removal rates, surface quality, and cost information for conventional machining processes.

1.1 Introduction

Laser machining belongs to the large family of *material removing* or *machining* processes. In order for the reader to understand the significance and limitations of laser machining in the overall context of manufacturing, we will review in the following the most common conventional machining processes, which will help clarify the position of laser machining and its applications in the manufacturing environment.

Manufacturing is *the transformation of materials into goods for the satisfaction of human needs*. To create a manufacturing system, one must choose from a number of processes that will *alter the form and shape or the physical properties of the materials*. In order to deal efficiently with the great number of options available, it helps to consider a set of criteria whose importance changes in different situations. These criteria are *cost, production rate, flexibility, and quality*. The consideration of these criteria will help in identifying optimum solutions for most manufacturing problems.

1.2 Conventional Machining Processes

The brief review below is designed to clarify the physical mechanisms involved in the different types of machining processes. The capabilities of each type of process with regard to cost, production rate, flexibility and quality are all directly related to the physics of the process.

1. *Forming or primary forming processes* create an original shape from a molten or gaseous state, or from solid particles. During primary forming processes, cohesion is usually created among particles.

2. *Deforming processes* convert the given shape of a solid to another shape without changing its mass or its material composition. Cohesion is maintained among particles.

3. *Removing processes* remove material during the process itself, destroying cohesion among particles.

4. *Joining processes* unite individual workpieces to make subassemblies or final products. This class of process, which includes additive processes such as filling and impregnating of workpieces, increases cohesion among particles.

5. *Processes that change material properties* purposely change the material properties of a workpiece in order to achieve desirable characteristics.

The most commonly used materials in industrial manufacturing processes (Table 1.1) can be classified in four categories: *metals, ceramics, polymers* and *composites*. The selection of which process to apply to a particular material is influenced by a number of factors which affect cost and the other criteria mentioned above. Probably the two most important factors (Fig. 1.1) are:

1. *The lot size of the parts to be made.* Small lot sizes require flexible processes, such as material removal processes, which are able to accommodate different geometric features, etc. Large lot sizes allow the use of primary forming or deforming processes, since tool cost for such processes is high, requiring a large lot size to offset costs.

2. *Physical properties of the material (i.e. melting point).* Metals have a relatively high melting point, and therefore are usually processed in solid form using material removal and deforming processes. Polymers and epoxy-based composites have a much lower melting point, allowing the use of primary forming processes, where the material is often in liquid state; however, secondary operations, usually machining, are often needed to achieve the required dimensional accuracy and surface quality. Because these materials often contain abrasive fibers or fillers, they often exhibit abrasive behavior when conventionally machined. Ceramics are usually brittle, making them difficult to process in solid form with conventional machining techniques. With ceramics, primary forming processes are used to create the basic shape of the workpiece, and secondary operations (usually machining) are used to create the final shape and surface quality.

	Forming	Deforming	Removing	Joining	Changing
Metals	xx	xx	xx	xx	xx
Ceramics	xx	--	x	x	--
Polymers	xx	x	x	x	--
Composites	xx	--	x	x	--

TABLE 1.1   Process Application on Different Materials (xx: widely used, x: seldom used, --: not used)

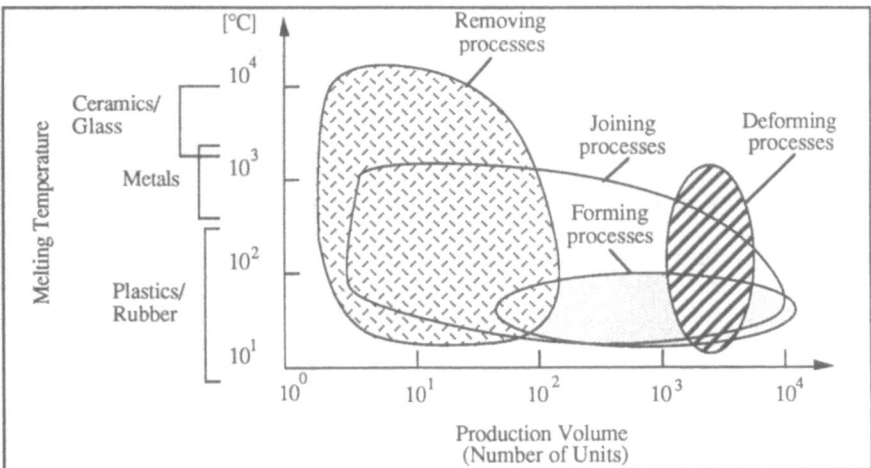

FIGURE 1.1   The Effect of Melting Point and Lot Size on Manufacturing Process Application

Because of their flexibility, material removal processes are particularly suitable for small lot sizes. Conventional machining processes are suitable for ductile materials such as metals and are much less applicable for brittle materials such as hardened metals, ceramics, and composites. Laser machining expands the application of machining processes to these difficult-to-machine materials.

The most important factor in removing processes is the material removal mechanism. There are four basic types of mechanisms:

Traditional Machining

- Mechanical: The mechanical stresses induced by a tool overcome the strength of the material.

Nontraditional Machining

- Thermal: Thermal energy provided by a heat source melts and/or vaporizes the volume of the material to be removed.
- Electrochemical: Electrochemical reactions induced by an electric field into an electrolyte destroy the atomic bonds of the material to be removed.
- Chemical: Chemical reactions destroy the atomic bonds of the material to be removed.

Choosing a removal mechanism requires consideration of a number of factors, the two most important being the hardness and/or abrasiveness of the material and the achievable material removal rate (Fig. 1.2). Traditional machining techniques span a wide variety of material removal rates; however, these rates drop significantly when the material hardness and/or abrasiveness is high. For such materials, *non-traditional machining processes*, based on thermal, electrochemical, and chemical removal mechanisms, are more practical.

Removing processes are characterized by their flexibility in terms of lot size, workpiece geometry and shape, and achievable quality. Particularly in traditional machining techniques, a multi-axis motion can be achieved between the tool and the workpiece, allowing the creation of a large variety of part sizes and shapes. Surface finish in traditional machining varies significantly; lower feed rates usually improve surface quality, but sacrifice material removal rate. Because of their flexibility, material removal processes based on the mechanical removal mechanism are the most widely used in industrial manufacturing. A summary of common material removal processes follows.

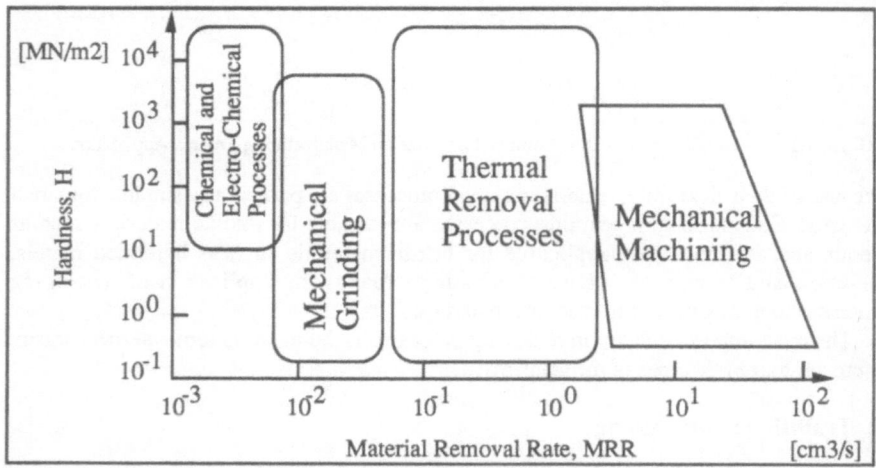

FIGURE 1.2   The Effect of Material Removal Rate and Hardness on the Application of Material Removal

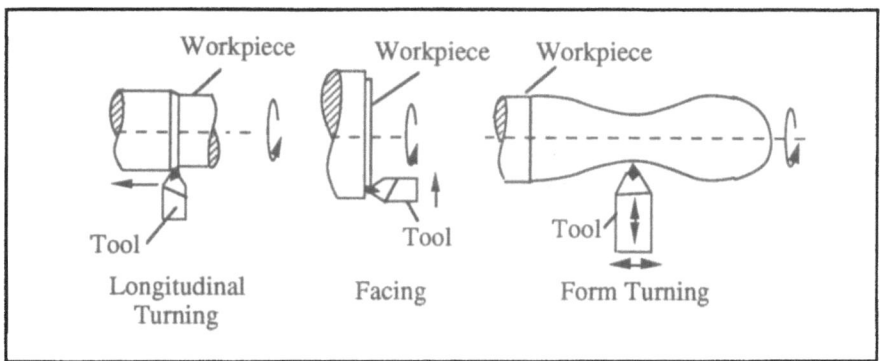

FIGURE 1.3   Common Turning Processes

Turning (Fig. 1.3), which can include longitudinal turning, facing, and form turning (depending on feed motion), is probably the most commonly used manufacturing process.

In *drilling,* the cutting action is not induced by rotational motion of the workpiece, but by the rotational motion of the tool (Fig. 1.4). Drilling includes twist drilling, boring, counter-sinking, reaming, and tapping. Due to the nature of the drilling process, chip formation occurs in a closed space. Drilling is, in general, a slow process with a relatively low production rate. In deep hole drilling, where the aspect ratio of the holes may be greater than 1:200, special consideration is given to the design of the tool, which has internal coolant channels that facilitate the ejection of chips from the deep holes. Machinery design must take into account the fact that the tools are very slender, creating stability problems.

Milling is similar to drilling, in that the main rotary cutting movement is produced by the tool. However, in milling, the feed motion is not in the axial direction of the cutting tool, as it was in drilling, but is vertical to the main axis of the tool. *Roller milling* and

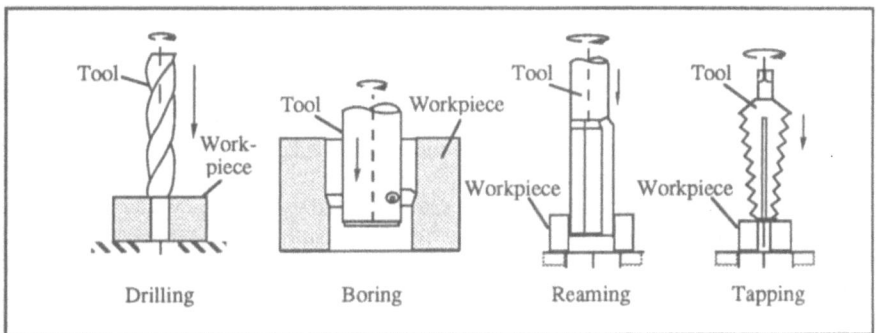

FIGURE 1.4   Common Drilling Processes

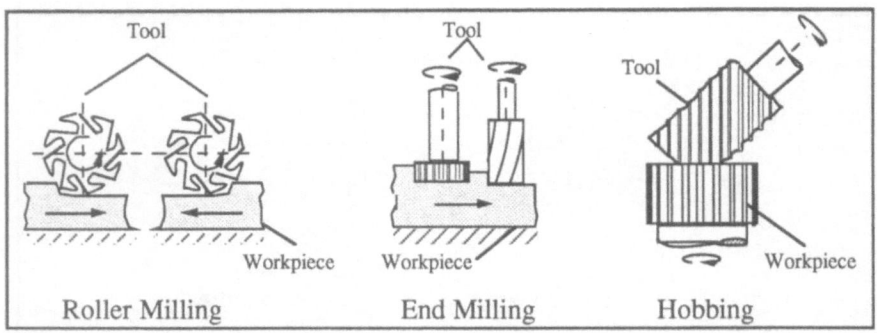

FIGURE 1.5   Common Milling Processes

end milling are general purpose processes (Fig. 1.5) that are widely used for producing flat surfaces in any type of configuration. *Hobbing* is common in gear making. Other varieties of the milling process include *form end milling*, which is important for the machine tool industry because it lends itself to milling slides and other machine tool parts; *die sinking*, which is used to make dies for deforming processes, and *gang milling*, where a variety of tools are mounted on the same rotating axis to make a complex shape.

The laser machining application that is most comparable with these traditional machining processes is *laser drilling*. However, if a single laser beam is used for turning or milling operations, the material to be removed must be entirely vaporized, or molten. Herein lie two major drawbacks to laser machining. First, despite laser machining's flexibility in terms of beam motion, it is often impossible to create three-dimensional shapes with lasers as easily as they can be produced using conventional turning or milling. Second, laser machining removes material "atom by atom," since the laser beam energy causes the material to change its phase from solid to liquid and/or vapor. Thus the energy efficiency of the laser machining process is much lower than conventional machining, where material is removed "chunk by chunk." These two drawbacks explain why laser machining has primarily been applied to drilling operations, particularly in drilling high aspect ratio holes with very small diameters, usually determined by the spot size of the laser beam (one or two orders of magnitude smaller than the smallest twist drill). There are no problems with tool stability, wear and breakage, and lasers do not need special tool or workpiece fixtures. Later in this text, the concept of three-dimensional laser machining using multiple beams will be discussed. This minimizes the aforementioned drawbacks of laser machining and allows laser beams to be used as tools for turning and/or milling operations.

Another process, which provides a competitive edge to laser machining capabilities, is *sheet metal cutting*, in which blanks are cut from large rolled sheets and further processed into desired shapes. Before a sheet metal part is made, a suitably sized blank is removed from a large, usually coiled sheet by *shearing*; this involves the relative movement of a punch and a die (Fig. 1.6), leading to the development of shear stresses higher than the strength of the workpiece material. Punching and blanking are based on the shearing process. Laser cutting can be used here instead of mechanical shearing, by

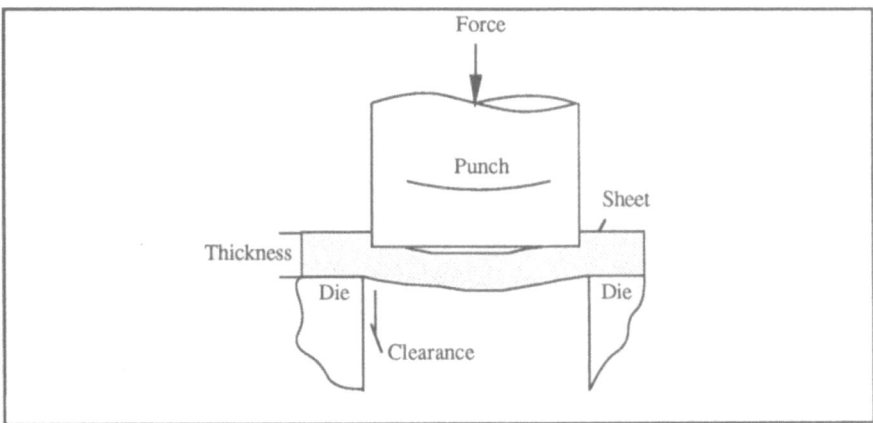

FIGURE 1.6   Metal Shearing With a Punch and Die

creating the desired shape through the relative motion between the metal sheet and the laser beam.

A large variety of shapes, from rotationally symmetrical to prismatic parts, can be produced by cutting with geometrically defined single cutting edge tools. The workpieces produced by these processes, however, are not always satisfactory in terms of surface quality and dimensional accuracy. To improve this, another set of material removal processes – those in which the cutting tool has a large number of cutting edges – are applied. These processes, which can be used with very hard workpiece materials, concentrate on the achievable surface quality and dimensional accuracy of the workpiece rather than the material removal rate. *Grinding* is the primary process in this category (Fig. 1.7). The grinding process sacrifices material removal rate to achieve high surface quality. Two other high-precision processes have the same general characteristics as grinding: *honing* and *lapping*. With honing and lapping, the surface finish can be improved, but not the dimensional accuracy of the workpiece, because the tools (e.g. honing stones in the case of honing) are not guided as they are applied to the workpiece (Fig. 1.8).

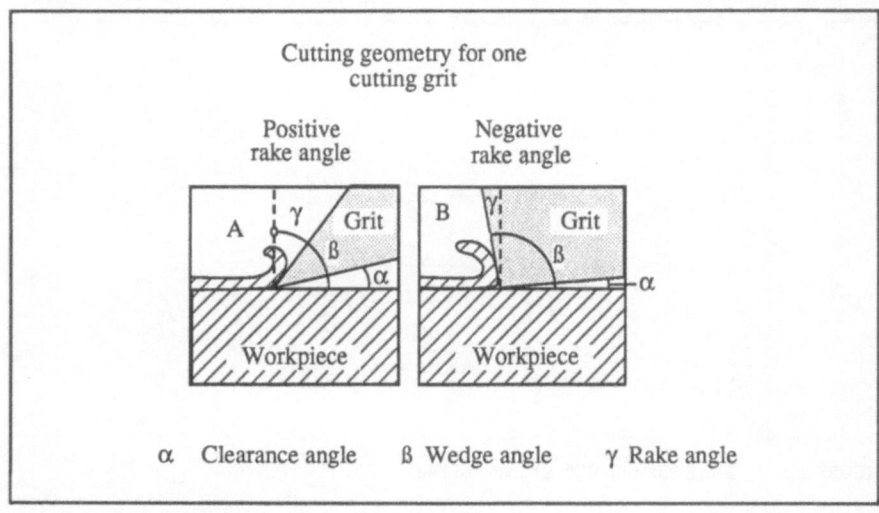

FIGURE 1.7　　　Basic Mechanism of the Grinding Process

Grinding	Honing	Lapping

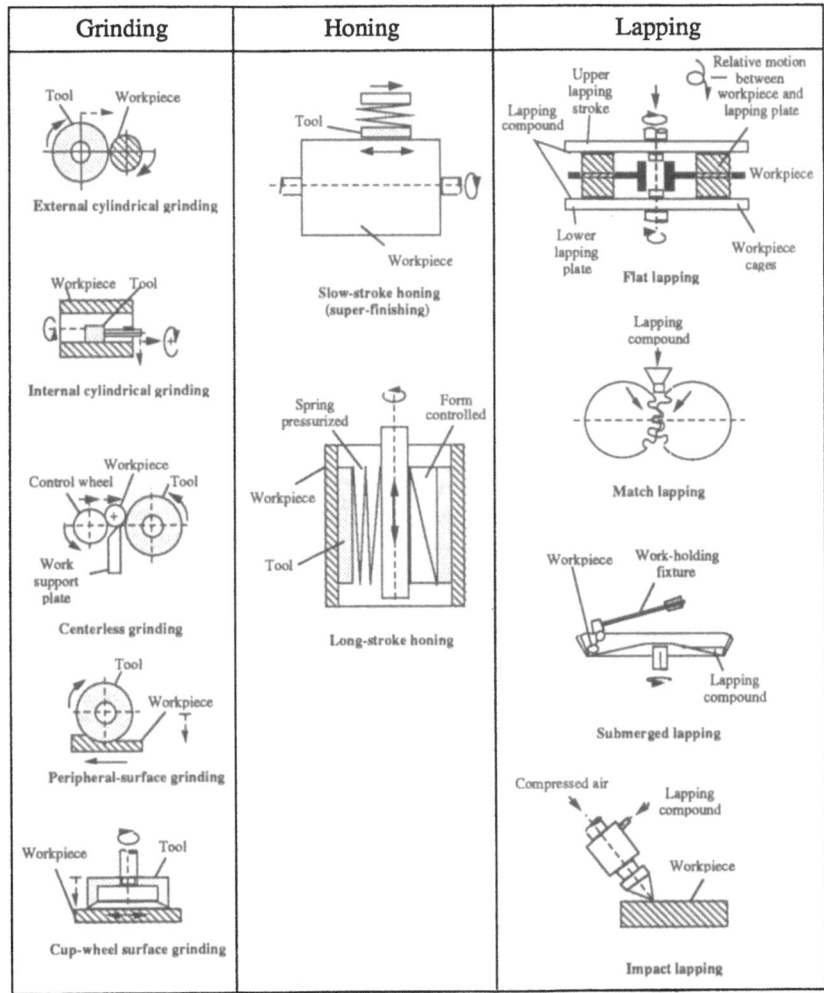

FIGURE 1.8   Basic Grinding, Honing and Lapping Processes

The cost of cutting operations has been widely investigated because of their widespread use in industrial practice (Tables 1.2 and 1.3). Typical material removal rates for laser machining range from 0.1 to 1 cm^3/min. This is less than the material removal rates achievable for turning and milling operations (1 to 50 cm^3/min), but compares favorably with mechanical drilling (0.001 to 0.01 cm^3/min) and grinding (10^{-4} cm^3/min or less) processes. In general, there is a tradeoff between production rate (cutting speed) and cost for any machining process (Fig. 1.9).

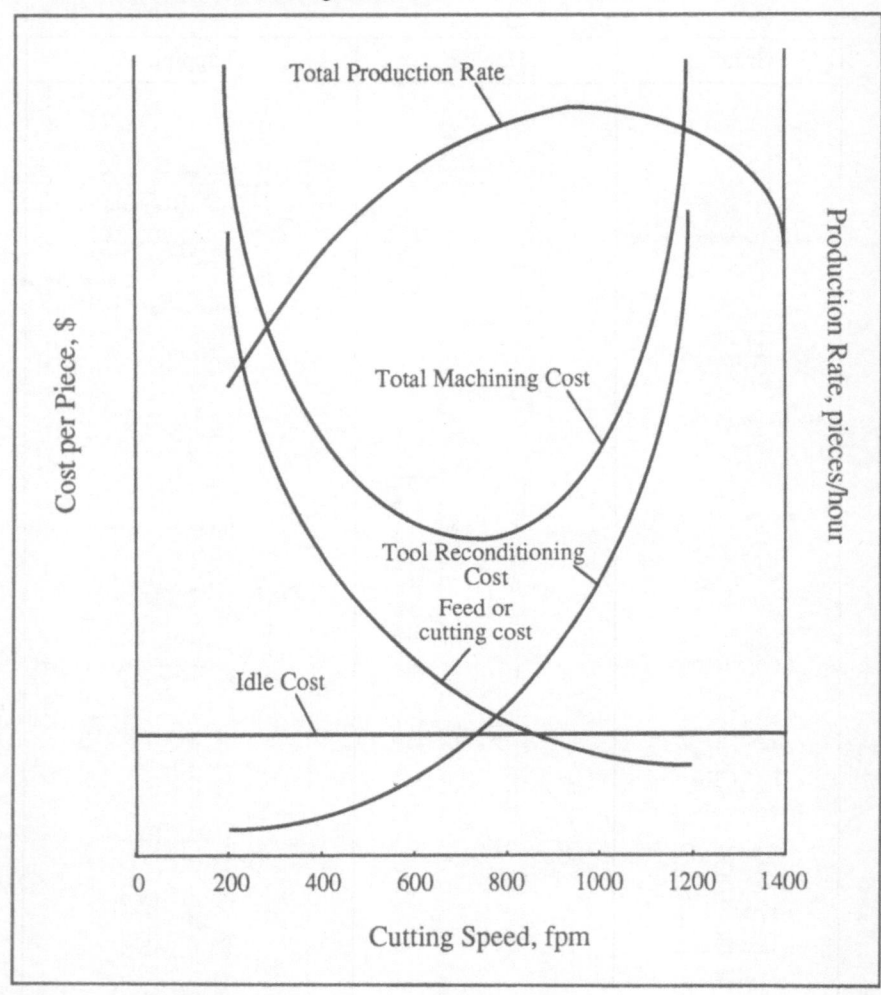

FIGURE 1.9   Machining Cost and Production Rate [7]

Turning

$$C = M\left[\frac{D(L+e)}{3.82f_r v} + \frac{R}{r} + \frac{DL}{3.82f_r vT} + t_i + \frac{DL\,t_d}{3.82f_r vT}\right]\left[\frac{C_p}{(K_t+1)} + Gt_s + \frac{G_{tb}}{k_2} + \frac{C_c}{k_1} + C_w + Gt_p\right]$$

Milling

$$C = M\left[\frac{D(L+e)}{3.82Zf_t v} + \frac{R}{r} + \frac{L}{ZT_t} + t_i + \frac{L\,t_d}{ZT_l}\right]\left[\frac{C_p}{(K_t+1)} + Gt_s + \frac{G_{tb}}{k_2} + \frac{ZC_c}{k_3} + C_w + Gt_p\right]$$

Drilling or Reaming

$$C = M\left[\frac{D(L+e)}{3.82f_r v} + \frac{R}{r} + \frac{L}{T_t} + t_i + \frac{L\,t_d}{T_l}\right]\left[\frac{C_p}{(K_t+1)} + Gt_s + Gt_p\right]$$

Tapping

$$C = M\left[\frac{mD(L+e)}{1.91v} + \frac{R}{r} + \frac{L}{T_t} + t_i + \frac{L\,t_d}{T_l}\right]\left[\frac{C_p}{(K_t+1)} + Gt_s + Gt_p\right]$$

Center Drilling or Chamfering

$$C = M\left[\frac{D(L+e)}{3.82f_r v} + \frac{R}{r} + \frac{U_c}{T_h} + t_i + \frac{U_c t_d}{T_h}\right]\left[\frac{C_p}{(K_t+1)} + Gt_s + Gt_p\right]$$

Handling and Setup

$$C = M\left[t_L + \frac{t_o}{N_L}\right]$$

$\frac{\$}{Min}$	Feeding Time	Rapid Traverse Time	Load & Unload Time	Setup Time	Cutter Index Time	Dull Tool Replacement Time	Tool Depreciation Cost	Tool Resharpening Cost	Rebrazing or Blade Reset Cost	Insert or Blade Cost	Grinding Wheel Cost	Tool Presetting Cost

TABLE 1.2   Machining Cost Relationships [7]

VARIABLE	DEFINITION	APPLIES TO OPERATION				
		Turn	Mill	Drill and Ream	Tap	Center Drill
C	Cost for machining one workpiece: $/workpiece	√	√	√	√	√
C_c	Cost of each insert or inserted blade: $/blade	√	√	No	No	No
C_p	Purchase cost of tool or cutter: $/cutter	√	√	√	√	√
C_w	Cost of grinding wheel for resharpening tool or cutter: $/cutter	√	√	No	No	No
d	Depth of cut: in	√	√	No	No	No
D	Dia. of work in turning, of tool in milling, drilling, reaming, tapping: in	√	√	√	√	√
e	Extra travel at feedrate (f_r or f_t) including approach, overtravel, and all positioning moves: in	√	√	√	√	√
f_r	Feed per revolution: in	√	No	√	√	√
f_t	Feed per tooth: in	√	√	√	√	√
g	Labor & overhead in tool reconditioning department: $/min	√	√	√	√	√
k_1	No. of times lathe tool, or milling cutter, or drift, or reamer or tap is resharpened before being discarded	√	√	√	√	√
k_2	No. of times lathe tool or milling cutter is resharpened before inserts or blades are rebrazed or reset					
k_3	No. of times blades (or inserts) are resharpened (or indexed) before blades (or inserts) are discarded	√	√	√	√	√
L	Length of workpiece in turning and milling or sum of length of all holes of same diameter in drilling, reaming, tapping: in	√	√	√	√	√
m	No. of threads per inch	√	√	√	√	√
M	Labor & overhead cost on lathe, milling or drilling machine: $/min					
n	Tool life exponent in Taylor's equation	√	√	√	√	√
N_L	No. of workpieces in lot	√	√	√	√	√
N_s	No. of pieces between sharpenings	√	√	√	√	√
P	Production rate per hour: workpieces/hour	√	√	√	√	√
r	Rapid traverse rate: in/min	√	√	√	√	√
R	Total rapid traverse distance for a tool or cutter on one part: in	√	√	√	√	√
S	Reference cutting speed for a tool life of T = 1 min: fpm	√	√	√	√	√
S_t	Reference cutting speed for a tool life of T_t = 1 min: fpm	√	√	√	√	√
t_b	Time to rebraze lathe tool or cutter teeth or reset blades: min	√	√	√	√	√
t_d	Time to replace dull cutter in tool changer storage unit: min	√	√	√	√	√
t_i	Time to index from one type cutter to another between operations (automatic or manual): min	√	√	√	√	√
t_L	Time to load and unload workpiece: min	√	√	√	√	√
t_m	Time (average) to complete one operation: min	√	√	√	√	√
t_o	Time to setup machine tool for operations: min	√	√	√	√	√
t_p	Time to preset tools away from machine (in toolroom): min	√	√	√	√	√
t_s	Time to resharpen lathe tool, milling cutter drill, reamer or tap: min/tool	√	√	√	√	√
T	Tool life measured in minutes to dull a lathe tool: min	√	√	√	√	√
T_h	No. of holes per resharpening	√	√	√	√	√
T_t	Tool life measured in inches travel of work or tool to dull a drill, reamer, tap or one milling cutter tooth, in	√	√	√	√	
u_c	No. of holes center drilled or chamfered in workpiece					
v	Cutting speed: fpm	√	√	√	√	√
w	Width of cut: in	√	√	√	√	√
Z	No. of teeth in milling cutter or no. of flutes in a tap	√	√	√	√	√

TABLE 1.3   Machining Rate Relationships [7]

Surface quality, and surface technology in general, are important aspects of material removal processes. A number of factors affect surface texture and surface integrity (Fig. 1.10). *Surface texture* describes surface roughness, macro effects such as imperfections, and geometric considerations such as tolerances. *Surface integrity* refers to micro-structure effects such as micro-cracks and residual stresses. *Surface effects* are caused by the process itself as well as by the workpiece material's properties; they have a direct influence on the mechanical properties of the workpiece and eventually on the reliability of the component. There are industrial standards used internationally to define surface quality characteristics; Table 1.4 summarizes the achievable surface roughness of the different processes, including commonly used processes such as drilling, milling, and grinding and primary forming processes such as investment casting and die casting. Table 1.4 demonstrates that processes based on thermal removal mechanisms, such as flame cutting or even laser machining, usually provide an inferior surface quality compared to processes based on mechanical removal mechanisms. However, for some materials such as glass, laser machining results in a polished kerf surface which is superior to the surface quality achievable for mechanical cutting. The table also shows that milling, turning, and grinding processes span a wide range of achievable surface quality corresponding to the material removal rate used, which explains their popularity in industrial practice.

In summary, laser machining provides a feasible production method for difficult-to-machine materials and special applications such as micromachining, while it has limitations in terms of material removal rate and surface quality when compared with traditional machining methods.

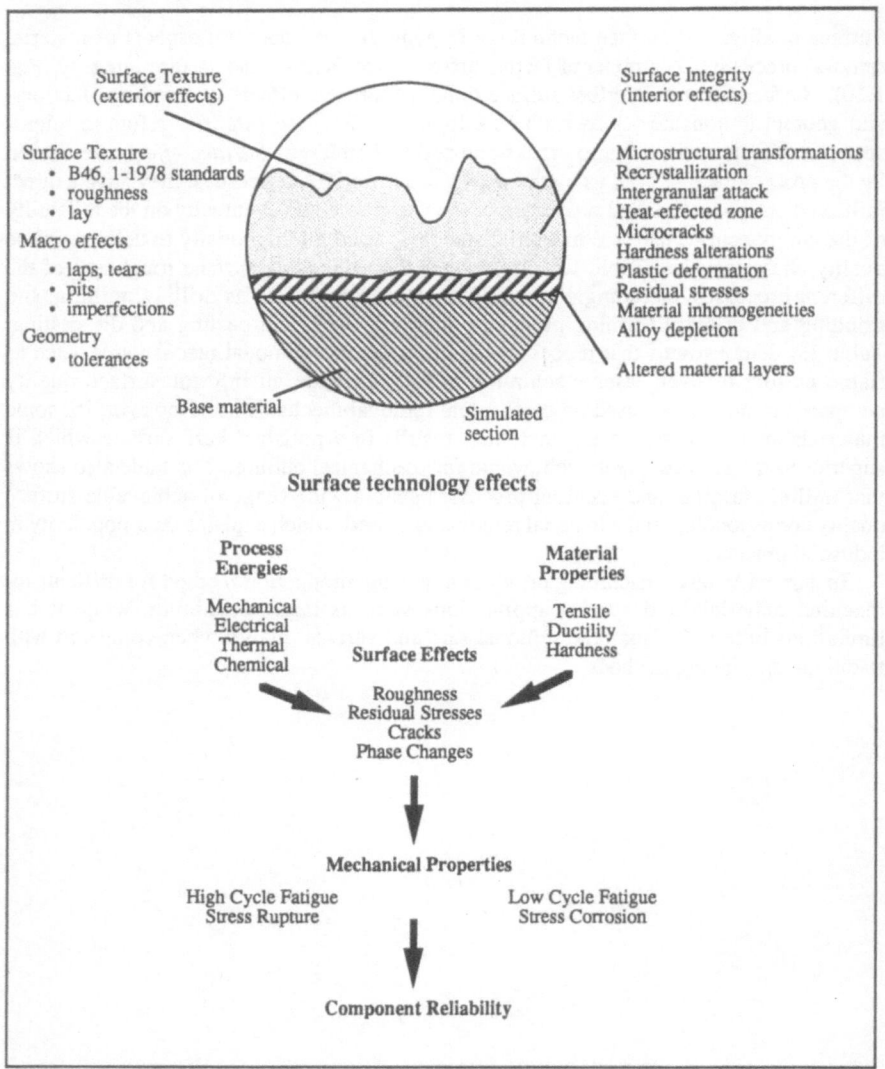

FIGURE 1.10   Machined Surface Characteristics [7]

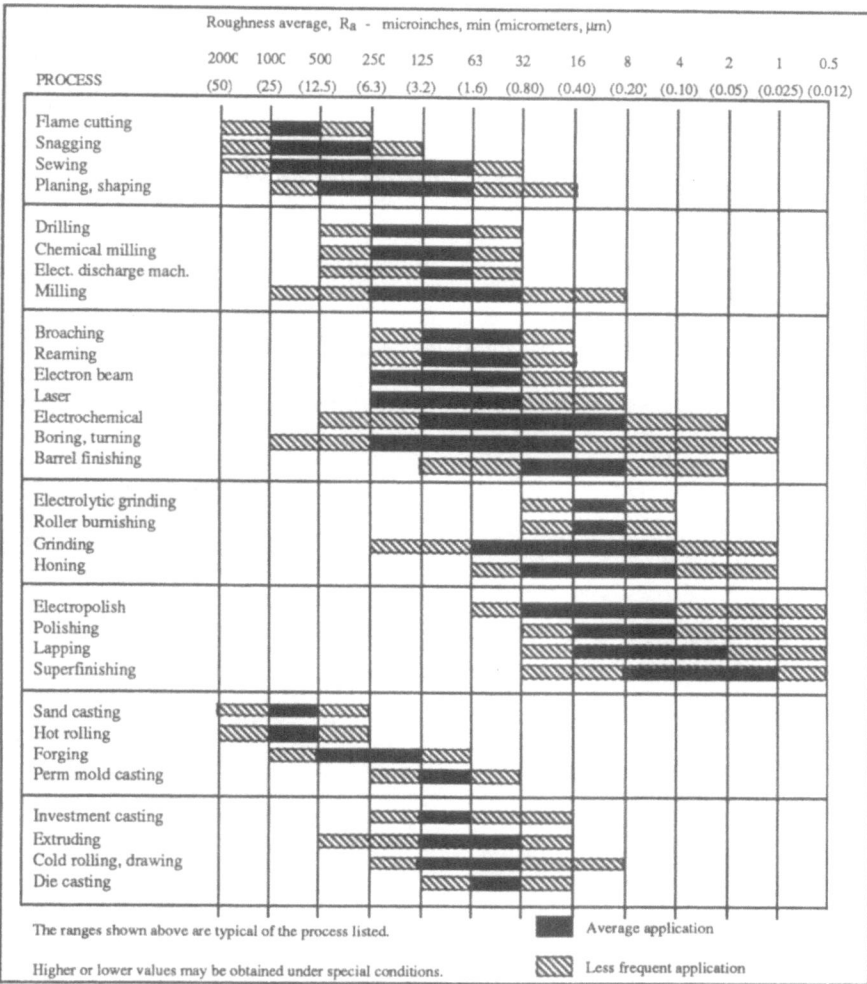

TABLE 1.4   Surface Quality for Different Machining Processes [7]

References

1.   Belforte, D., *The Industrial Laser Annual Handbook*, Pennwell Publishing, Tulsa, OK, 1987.

2.   Kalpakjan, S., *Manufacturing Engineering and Technology*, Addison-Wesley, Cambridge, MA, 1989.

3.   König, W., *Fertigungsverfahren (Manufacturing Processes)*. VDI-Verlag, Düsseldorf, W. Germany, 1984.

4.   Lange, K., *Handbook of Metalforming*, McGraw Hill, New York, 1985.

5.   Schey, J.A, *Introduction to Manufacturing Processes*, McGraw Hill, New York, 1974.

6.   Weck, M., *Werkzeugmaschinen*, VDI-Verlag, Düsseldorf, W. Germany, 1979.

7.   *Machinability Data Handbook*, Metcut Research Associates Inc., Cincinnati Ohio, 1980.

8.   *Tool and Manufacturing Engineers Handbook*, SME, Dearborn, Michigan, 1983.

2
Lasers for Machining

The laser is clearly the most critical component of any laser machining system. This chapter provides an overview of the operation of the various types of lasers used in machining. First, the basic mechanisms required to produce laser light, such as stimulation, amplification, and population inversion are explained. The unique properties of laser light, including monochromacity, coherence, diffraction, and radiance are also described. Lasers can be classified by the type of lasing medium used: solid, liquid or gas. The beam each type of laser produces has a characteristic wavelength and power range. The most common types of lasers used for machining are CO_2, Nd:YAG, and excimer lasers. Possible configurations for these three laser types are detailed.

2.1 Lasers and Light: A Historical Perspective

Any discussion of lasers and laser light must begin with an introduction to ordinary light. In 1704, Newton characterized light as a stream of particles because of its tendency to travel in a straight line [7]. However, he also recognized that light possessed certain wavelike properties. Nevertheless, Newton's description of light as a stream of particles was accepted for the next century. Early in the nineteenth century, new evidence again showed that light had wavelike properties. In 1803, an experiment by Young, a London physician, provided conclusive evidence. Monochromatic light was passed through two pinholes, resulting in an interference pattern resembling waves in water or pulses of sound. Concurrently, Fresnel and Arago correctly interpreted an experiment by Huygens, who had shown early in the seventeenth century that light transmitted through calcite crystals was polarized, convincing him that light was composed of waves. Fresnel and Arago re-examined this experiment and explained that light waves were transverse waves oscillating perpendicularly to the direction of propagation and were not longitudinal waves, as Huygens had postulated.

The discovery of polarity complimented the electromagnetic theory proposed by Maxwell, who described light as rapid variations in the electromagnetic field due to the oscillation of charged particles [7]. Thus, light was classified as merely one form of the radiant energy which ranged in form from radio waves to X rays; these forms were all phenomena relating to electromagnetism.

The interpretation of light as wavelike was challenged with the advances of physics in the twentieth century. In 1900, Planck began an investigation of emissions from hot bodies [7]. According to electromagnetic theory, the intensity of emissions at a particular frequency should be proportional to the square of the emitted frequency. This would mean that an almost infinite amount of energy would be radiated at higher frequencies; however, markedly different results were obtained in actual experimentation. Planck derived an empirical formula to explain the relationship between intensity and the frequency of the emitted radiation; he reasoned that light must exchange energy with matter in quanta or "packets". His equation showed that the energy of each packet was related to the frequency of the light by a constant (h = 6.63 x 10^{-34} joule-sec), a fundamental constant of nature now known as *Planck's constant*.

In 1905, Einstein further extended the quantum concept of light. He noted that electromagnetic theory failed to explain the photoelectric effect, in which negatively charged metal plates lost their charge when exposed to radiation of sufficient energy [7]. The electrons in the metal absorbed radiant energy and attained enough excitation to leave the plate. If the energy of a light wave were distributed over its entire length, an amplitude of radiation, which would be below the threshold energy level for ejecting an electron, should have existed, and the metal would have required a period of exposure to 'charge up' before emitting any electrons. However, no such amplitude or delay was detected, and Einstein maintained that the quantization of light was the only possible explanation.

Light, then, possesses certain wavelike properties, such as polarization and interference, but it can also behave as if composed of individual particles, called photons, which possess a discrete amount of energy or quanta. In the 1920's, an understanding of these phenomena was first developed, and the field of *quantum mechanics* was formed. This theory describes light as composed of particles which do not obey deterministic laws of motion, as in classical mechanics. Thus, the motion of a particle in a known force field cannot be determined exactly, but rather its wavelike behavior is an expression of the probabilities that govern its motion. This can be dramatically demonstrated by an interference experiment where a single photon is in motion on the pickup screen. The pattern of the first photons appears to be random, but with long exposures, the expected interference pattern develops.

Although special in many ways, laser light does comply with the basic laws of quantum mechanics and exhibits both wavelike and particle-like properties. LASER is an acronym for Light Amplification by Stimulated Emission of Radiation. Laser light differs from ordinary light in that *it consists of photons that are all at the same frequency and phase (coherence)*. A laser's ability to produce *coherent* light is based upon the principle that photons of light can stimulate the electrons of atoms so they emit photons of the exact same frequency [17]. This occurs when a photon passes close to the electron, and can be explained using quantum mechanics. The possibility of stimulation was first postulated by Einstein in 1917, but the first working laser was not built until almost half a century later.

2.2 Basic Mechanisms in Lasers

Since the central component of any laser machining system is the laser itself, it is necessary to understand the mechanisms which cause laser light when discussing laser machining. Lasers convert electrical energy into a high energy density beam of light through *stimulation* and *amplification* [17]. *Stimulation* (Fig. 2.1) occurs when electrons in the lasing medium are excited by an external source such as an electrical arc or flash lamp, resulting in the emission of photons [3, 15]. The energy required to raise an electron from one energy state to another is provided by an excitation process, or *pumping*. This is achieved by the lasing medium's absorption of energy from mechanical, chemical, electrical, or light sources. The rate of power input to the excitation process must exceed the output power of the laser, since there are many losses associated with the lasing process. The lasing medium typically contains ions, atoms, or molecules whose electrons are conducive to changes in energy level. According to quantum mechanics, atoms or molecules in the lasing medium have discrete electron energy levels. Laser light is created by the transition from a higher to a lower energy level, and the wavelength produced is a characteristic of the lasing medium. At the beginning of the lasing process, photon emissions are random in nature. As each photon stimulates other excited electrons to emit photons, however, the new photons will have similar wavelength, direction, and phase characteristics as the initial photon. Eventually, a stream of photons with identical wavelength, direction, and phase will be produced.

The *amplification* of light in a laser is accomplished by an optical resonator, which is composed of a cavity with the lasing medium set between two high-precision, aligned mirrors [3, 15]. One mirror is fully reflective, and the other is partially transmissive to allow for the beam output. The mirrors channel the light back into the lasing medium; as the photons pass back and forth through the lasing medium, they stimulate more and more emissions. Photons that are not aligned with the resonator are not redirected by the mirrors to stimulate more emissions, so that the cavity will only amplify those photons with the proper orientation, and a coherent beam develops quickly (Fig. 2.2).

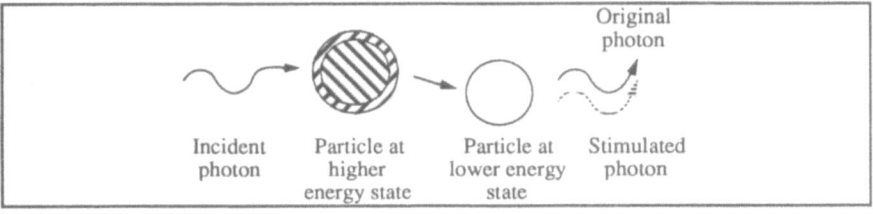

FIGURE 2.1   Stimulation

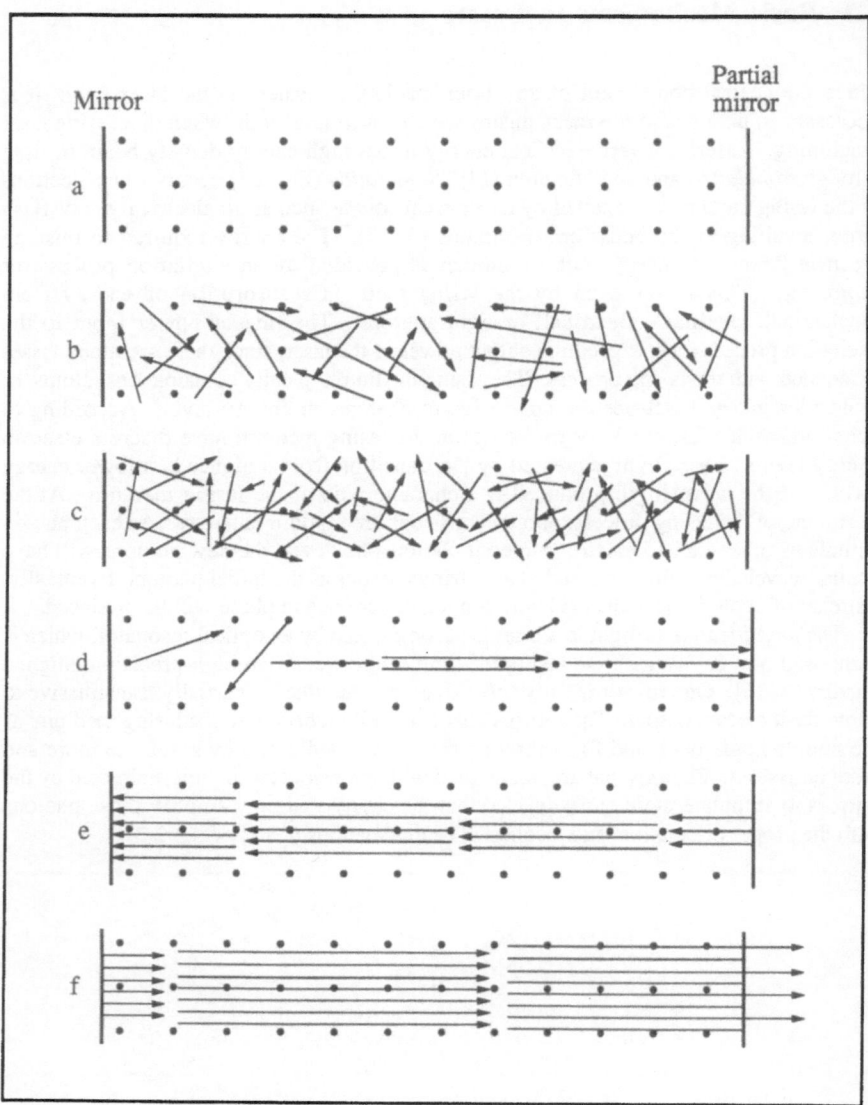

FIGURE 2.2   Amplification States (a) Laser Off, (b)and (c) Initial Random States, (d) Initial
Stimulation, (e) Amplification and (f) Coherent Beam

Population inversion [15] is another necessary condition for the lasing process (Fig.
2.3). When the lasing medium or lasant is in equilibrium, the population of electrons at
any energy state is determined by the Boltzmann equation. For a medium with two
energy states, the relationship between energy and electron population is:

$$\frac{N_2}{N_1} = \exp - \left(\frac{E_2 - E_1}{kT} \right)$$

(2-1)

where N_1 and N_2 are the number of electrons at Energy States 1 and 2 respectively, E_1 and E_2 are the energy values for States 1 and 2, T is the absolute temperature of the medium, and k is Boltzmann's constant. Since the ratio of population of any two energy states is related to the exponential of the inverse of the difference between the energy states, higher energy states always possess a smaller population than lower energy states under equilibrium conditions. The goal of choosing a lasing medium and an excitation method is to induce a non-equilibrium state that contains more high energy state than lower energy state electrons. Under the condition of a population inversion, sustained lasing action is possible because statistically more electrons are available to provide stimulated emissions than there are ground state electrons, which have the tendency to absorb the emitted photons (State 3 in Fig. 2.3). Theoretically, only a slight population inversion is required to achieve lasing; however, with other energy losses also occurring, there must often be several excited electron states for every equilibrium state electron. The product of the actions of stimulation, amplification and population inversion is to produce a stream of photons with common characteristics, namely a laser beam. This laser beam possesses temporal, spatial and energy properties which are different than diffuse light. These properties make the laser beam useful for many applications in communications, measurement, and materials processing.

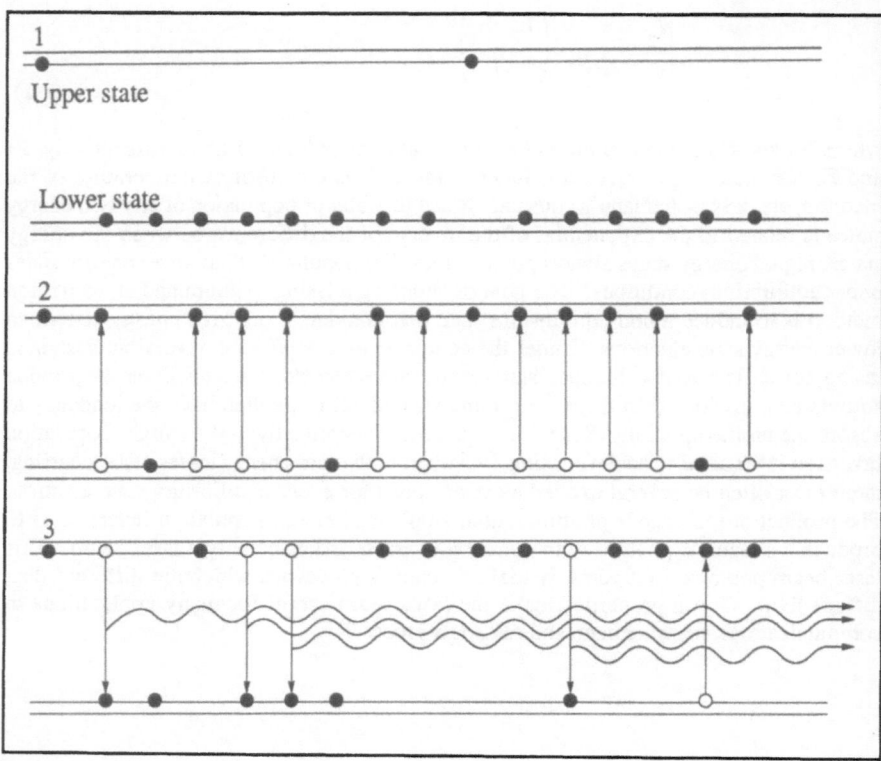

FIGURE 2.3   Population Inversion

The first type of laser developed was the ruby laser, invented in 1957 by Townes and Shawlow. The ruby laser consists of a ruby lasing rod which has mirrored ends that form the optical resonator (Fig. 2.4) and a helical flash lamp which surrounds the ruby rod and provides excitation [17]. The ruby lasing medium is composed of aluminum oxide (Al_2O_3) with a small amount of dissolved chromium oxide (Cr_2O_3). The Cr^{3+} ions in chromium oxide give ruby lasers their distinctive red color and also allow for the generation of laser emissions (Fig. 2.5). In the Cr^{3+}:Al_2O_3 system, the initial or *ground state* of the Cr^{3+} ion when the laser is turned on is known as $^4A_2$. The helical flash lamp is used to raise or *pump* the electrons in the Cr^{3+} ions to the higher energy states of $^4T_2$ and $^4T_1$. Then, the energy levels for these excited electrons are lowered through a process which produces no radiation to an intermediate ^2E state, which is composed of two sublevels called $2\bar{A}$ and $\bar{E}$. The emission from the $\bar{E}$ sublevel to the ground state produces the characteristic 694.3 nm laser light of a ruby laser.

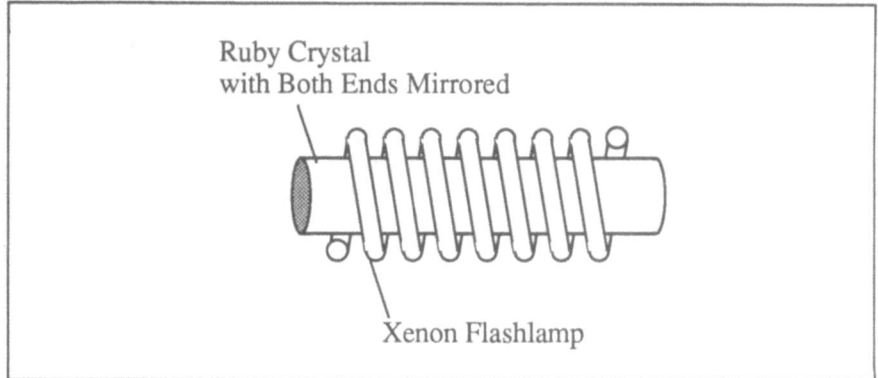

FIGURE 2.4   Ruby Laser

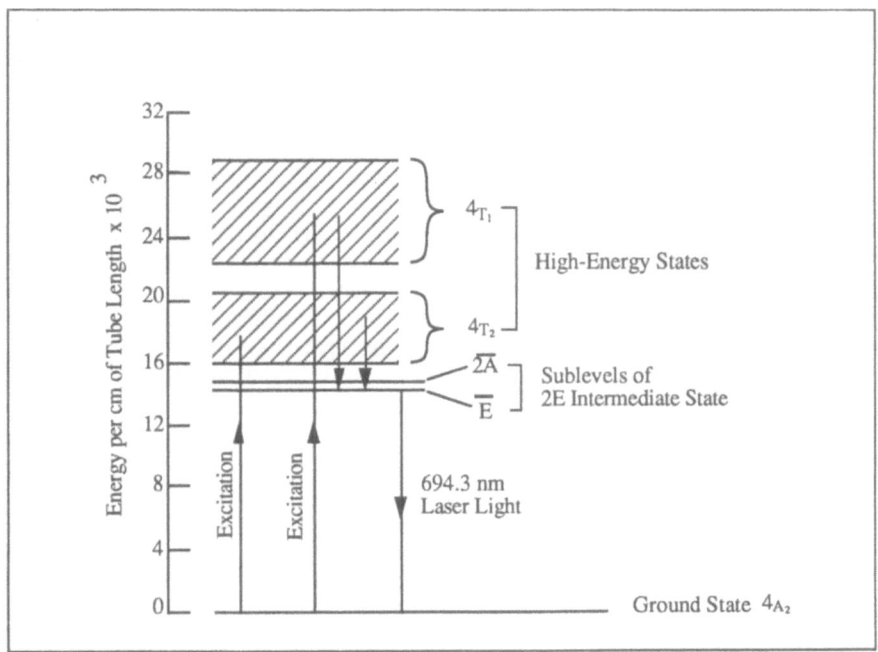

FIGURE 2.5   Chromium Ion Energy Levels

Ruby lasers are typically operated with a pulsed emission of the laser beam in order to obtain a high energy density because the energy efficiency ($\sim$0.1%) is too low for uninterrupted beam output. This creates leading edge spikes in the pulse output, because the stimulation of emissions rapidly depletes the stores of electrons in the ^2E state. In addition, the pumping of electrons to higher energy states is slow; therefore, pulse repetition rates are slow. The rate and efficiency of energy transfers to and from the excited electron energy states greatly influences the operating characteristics of any laser.

2.3   Laser Light: Properties

Simply illuminating a piece of paper can illustrate the inherent differences between laser and ordinary light. In ordinary light, a sheet of paper appears as a flat, smooth surface. The reflection of the light appears to be identical over the entire surface of the sheet, and the image does not vary greatly with changes in viewing angle. Under laser light, the paper will have a small illuminated area with a grainy appearance. This small illuminated area is due to the small divergence properties of the laser beam. The energy of the ordinary light source is spread over the entire piece of paper and most likely over the rest of the room. Laser light does not spread or diverge, and the spot on the paper is not much larger than the diameter of the beam at the output of the laser.

The grainy appearance of the paper under laser light is composed of contrasting regions of darkness and light, and the grainy pattern changes as the viewing angle is altered. This change is due to the scattering of the laser light by microscopic irregularities within the paper surface. Bright points appear in areas of constructive interference, while dark areas are created in areas of destructive interference. The uniform wavefronts of the laser beam make these interference patterns possible. In ordinary light, the wavefronts are unordered and produce chaotic interference patterns that have no discernible patterns.

The unique properties of laser light can be quantified by examining the optical properties of *monochromaticity, coherence, diffraction,* and *radiance.*

High *monochromaticity* implies that the range of frequencies emitted by the light source is small; this is often evaluated by measuring the spectral line width [23]. Laser light typically contains a single or a few spectral lines of very narrow widths (Fig. 2.6). Ordinary light sources have multiple wide lines. Monochromacity is most important in applications such as interferometry, holography, velocimetry, isotope separation, and communications in which the frequency content of the laser beam is of great importance. It is generally not a critical factor in laser machining.

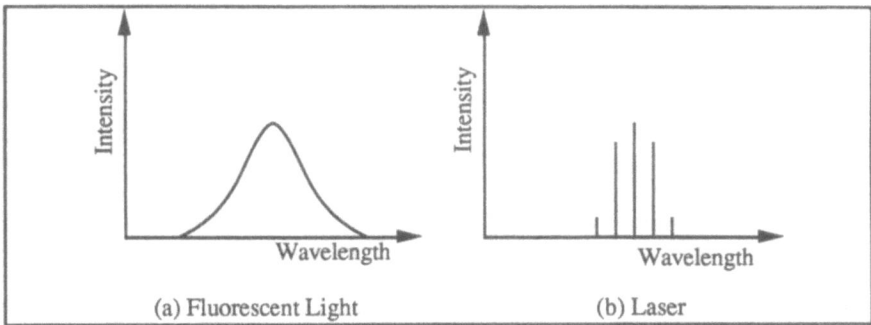

FIGURE 2.6   Monochromaticity

Spatial and temporal *coherence* refers to the relationship between the electronic and magnetic components of an electromagnetic wave (Fig. 2.7) [17].   When these components are all aligned, the beam is said to be coherent. Spatial coherence refers to the correlation of phases at different points in space at a single moment in time; temporal coherence refers to the correlation of phases at a single point in space over a period of time. An analogy to coherence can be made with a marching band. Looking at a row marching at a given moment, the members should all be aligned shoulder to shoulder; this is spatial coherence. Looking at a single column for a minute, the spacing between the members should be consistent; this is temporal coherence. An ordinary light source would more resemble the people at the intersection of several hallways in a building. The motions recorded for any area or point in time would be random.   Since monochromaticity is required for good coherence, the applications that are critically dependent on coherence are similar: interferometry, holography, velocimetry, and communications.

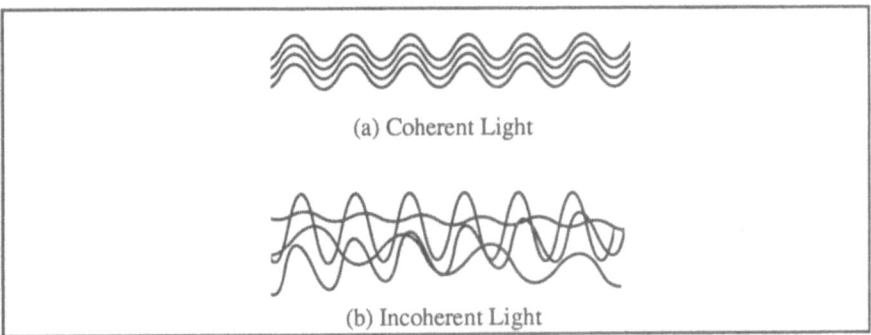

FIGURE 2.7   Coherence

Diffraction is the phenomenon by which light bends around sharp-edged objects [17]. When ordinary light is projected over a distance, a substantial amount of diffraction or scattering of light occurs, resulting in a loss of light intensity as the distance from the light source increases. One advantage of laser beams over ordinary light is that lasers produce beams with very limited diffraction. Low-diffraction light of this type can often be described as a *collimated beam*. Therefore, a laser beam can be projected over a distance with little divergence of the beam or loss in beam intensity. Typical beam divergence angles range from 0.2 milliradians for HeNe lasers to ten milliradians for Nd:YAG lasers. This property of laser light produces a directional energy source which can focus a large amount of luminous energy on a small area. In machining applications, this directional characteristic allows precise control of the laser beam direction relative to workpiece position. For the focussing of laser light, the diffraction effect is related to the ratio between focal length and aperture or beam diameter.

The *radiance* of a light source is the amount of power per unit area emitted by the light source for a given solid angle (Fig. 2.8) [15]. Typical units of radiance are Watts per square meter per steradian. The solid angle can be thought of as the cone through which the light passes. Lasers have high output powers for the small areas which they use to emit the beam. In addition, the beams possess low divergence, which causes them to be transmitted over a small solid angle. Thus, laser light sources possess extremely high radiance. A typical helium-neon laser with a one mW output possesses a radiance two orders of magnitude greater than that of the surface of the sun, which has a power output of 4×10^{26} W. The radiance of a source cannot be increased by optical manipulation; the high radiance capabilities of laser beams are influenced by the design of the laser cavity.

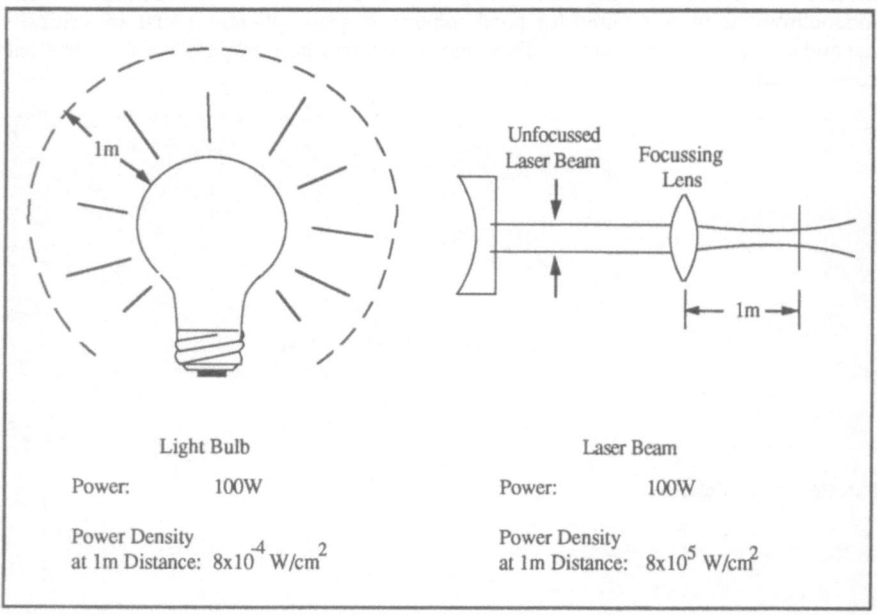

FIGURE 2.8   Radiance

While high monochromaticity, high coherence, and low diffraction individually are not critical factors for laser machining, the combination of these characteristics make the laser beam a powerful tool for machining. High power and energy density can be achieved on the surface of the workpiece, which leads to quick phase change of the workpiece material and to subsequent material removal.

2.4 Laser Types

Lasers can be categorized most easily according to their lasing mediums, which are divided into three basic categories as defined by the state of the lasing material: *gas*, *liquid*, or *solid*. Furthermore, all laser types operate in one of two temporal modes: *continuous wave* (CW) and *pulsed* modes. In CW mode, the laser beam is emitted without interruption. In pulsed mode, the laser beam is emitted periodically.

2.4.1 Gas Lasers

Gas lasers (Fig. 2.9) are further divided into three subgroups based on the composition of the lasing medium: neutral atom, ion, or molecular [19]. The helium-neon (HeNe) laser is a typical neutral atom gas laser and is the most popular visible light laser; it can be tuned from infrared to various visible frequencies, with the most common being red at a wavelength of $0.6328\mu m$. The actual lasing occurs during an electron transition in the neon (Ne) atom, but the presence of helium (He) is essential since it is excited first by an electrical discharge, and then its energy is transferred to the Ne atom through kinetic interactions. Excitation is provided by DC electrical discharge in a low pressure discharge tube. There are four possible high energy states that electrons in the lasing medium can achieve upon excitation. Since the lasing action has an energy efficiency of 0.1% or less, the laser beam energy is approximately three orders of magnitude lower than the electrical discharge energy. Emissions typically have a Gaussian energy intensity distribution with a maximum power of 50 mW. Areas of applications for neutral atom gas lasers include holography, scanning, alignment, measurement, vision, and optical-fiber communications, which all require a low-powered source of directional energy.

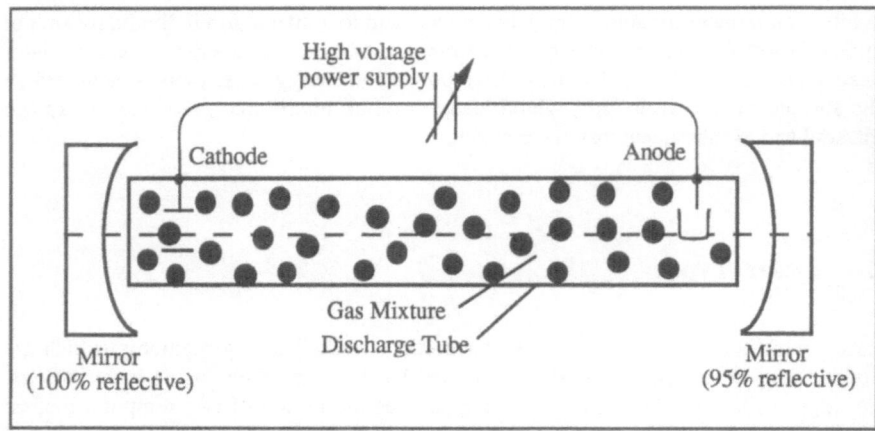

FIGURE 2.9   General Schematic of a Gas Laser

Ion gas lasers use an ionized gas such as argon (Ar), krypton (Kr), and xenon (Xe) as a lasing medium to produce laser beams with wavelengths ranging from 0.5 to 1.0 μm [15]. Excitation is initiated by an electrical discharge, which is achieved in two stages: first the gas is ionized, and then the electrons are brought to the excited state. The resulting laser beam can achieve power levels of up to several Watts. Typical applications include surgical applications and spectroscopy, which both require higher power levels than those achievable with neutral atom lasers.

Molecular lasers use gas molecules as the lasing medium, whereby the molecules are excited and the vibrational mode of the molecules change [15]; these transitions between different vibrational modes produce photons. Carbon monoxide (CO), hydrogen fluoride (HF), and carbon dioxide (CO_2) lasers are examples of this type, which is by far the most common type used for laser machining and which will be discussed in detail.

The *carbon dioxide laser* emits light at a wavelength of 10.6 μm in the far infrared region of the electromagnetic spectrum [10]. The lasing medium is a combination of the gases carbon dioxide, nitrogen, and helium. The lasing mechanism in a carbon dioxide laser differs from the solid or ion lasers. Excitation of the carbon dioxide is achieved by increasing the vibrational energy of the molecule. The actual pumping is achieved by an AC or DC electrical discharge. Only a small percentage of the carbon dioxide is excited directly by the electrical discharge; most of the electrical energy is absorbed by nitrogen gas. The vast majority of carbon dioxide molecules achieve excitation by colliding with the nitrogen molecules, which transfer their added vibrational energy (Figure 2.10). This is an extremely efficient and selective energy transfer because nitrogen is a dimer with one vibrational mode and only one excitation state (V=1 state). This excited state is very close to the required excitation state of the carbon dioxide (State 001). Once excitation is achieved in the CO_2 molecule, the electrons can release energy through collisions with other CO_2 molecules still in the ground state; these collisions lower the excited electrons to intermediate Energy States 100, 020, and 010. Of these energy changes, only the transformation of electrons from State 001 to State 100 releases energy in the form of 10.6μm laser light.

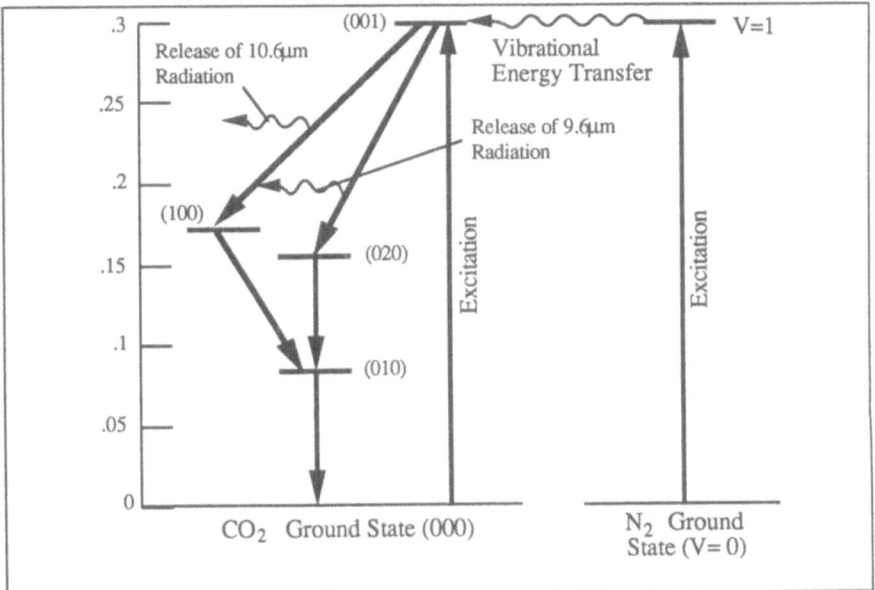

FIGURE 2.10   Carbon Dioxide Energy Levels

The remaining energy between the intermediate states and the ground state is lost through kinetic energy transfer, which generates heat instead of light. For CO_2 molecules, the rate of energy release through heat is much lower than energy release through light, so the energy efficiency for producing a laser beam is high compared with other lasing materials. In comparison, helium has a very high thermal diffusivity; therefore, with its addition to the lasing gas mixture, the rate of energy release through heating is extremely high. This combination of lasing interactions makes the carbon dioxide laser suitable for industrial applications both in terms of the energy efficiency (up to 10%) and the high output beam powers achievable.

Several different geometries of electrical discharge and gas flow are available. This contributes to the wide range of laser beam power levels which are available, from several Watts to 15 kW. High power carbon dioxide lasers typically operate with mixed energy intensity distributions; the beam profile may vary with time between several possible configurations [10]. Medium and low-powered CO_2 lasers are more stable and maintain a Gaussian intensity distribution. CO_2 lasers are widely used in industry for applications in laser machining, welding, and heat treating.

The properties of a carbon dioxide laser are primarily determined by the method of gas flow [10]. The three main types of gas flow are *sealed discharge tube*, *axial flow*, and *traverse* or *cross flow*. The flow method determines how fast the post-stimulation carbon dioxide gas can be removed from the optical cavity so that new ground state carbon dioxide gas can be introduced for excitation and stimulation.

The *sealed discharge laser* contains a fixed lasing gas mixture sealed in the laser cavity; therefore, it does not require a gas supply or gas handling system. However, because there is no gas flow, output powers are limited to about 50 W (since used CO_2 cannot be discharged and replenished with new CO_2), and the lifetime of the laser is limited by the disassociation of the carbon dioxide into oxygen and carbon monoxide.

The second type of CO_2 laser is the *axial flow laser*. This is the most widely used type of CO_2 laser, in which the gas flows along the axis of the optical cavity (Fig. 2.11) [1]. Axial flow allows the depleted gas to be replaced by new gas. Laser beam powers of up to 4 kW of continuous wave output can be achieved. Laser beams with pure Gaussian intensity profiles can be generated for power levels up to about one kW, while beams with power above one kW generally have a mixed mode output containing two or more different intensity profiles. Since the used gas is either exhausted immediately or is reused after removing any contaminants, this system requires a constant supply of gas and a gas handling system. Axial flow lasers can be further classified by the speed of gas flow, namely *low-speed flow* and *high-speed flow* lasers. Low-speed flow lasers attain power outputs of about 50 W to 70 W for each meter of cavity length. To produce a compact package, the optical cavity is folded so that longer discharge lengths can be obtained in a smaller assembly. The low flow speed causes the laser to heat up considerably, and the relatively low conductivity of the gas mixture limits the bore size of the lasing tube. Heating of the resonator cavity causes distortions in the resonator optics due to thermal expansion; these distortions affect the beam intensity profile and beam stability. However, with external cooling for the resonator and optics, beam outputs with good adjustability and stability can be generated. High-speed flow models typically have a gas velocity of 60 m/s. Thus, the carbon dioxide molecule only has time for one excitation/stimulation cycle before exiting the optical cavity. Typical outputs are 600 W per meter of cavity length, with total laser beam outputs available up to 6 kW. Due to the convective cooling of the high-speed flow, the thermal distortions in the resonator optics are minimized and larger bore diameters can be used to produce intensity profiles which are nearly Gaussian in shape.

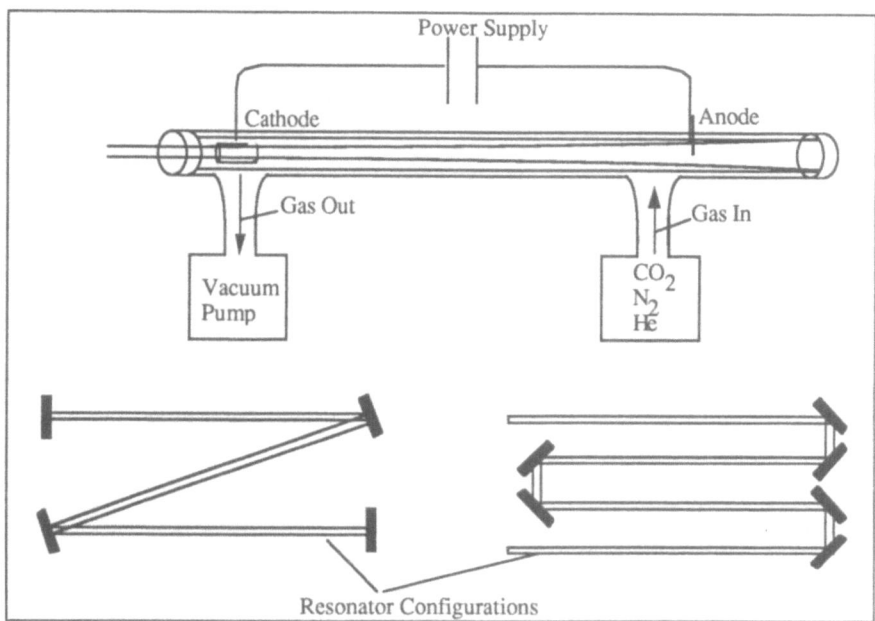

FIGURE 2.11   Axial Flow CO_2 Laser

The third type of CO_2 laser is the *transverse flow laser*, in which the gas mixture of He, N_2 and CO_2 flows perpendicular to the optical cavity axis (Fig. 2.12) [3]. The gas flow is maintained by a blower and is cooled by a heat exchanger. The electrodes are arranged on either side of the cavity, and relatively small voltages are required in order to maintain high current discharge due to the small separation distances. The beam is usually reflected or folded several times within the cavity, creating a compact design which can achieve outputs of one to 10 kW per meter of cavity length and total outputs greater than 15 kW. However, beam quality deteriorates at higher power levels and beam output can shift between several possible beam intensity profiles.

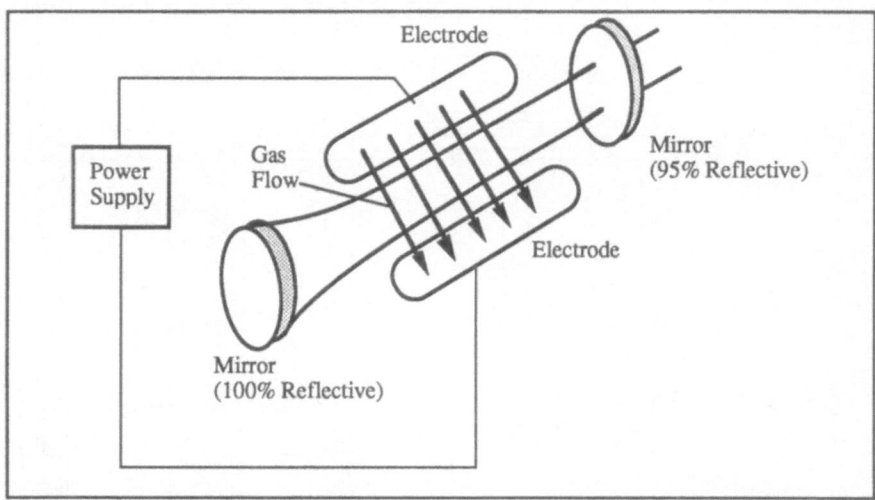

FIGURE 2.12   Traverse Flow CO_2 Laser

For all types of gas lasers, the electrical discharge is either AC, DC, or AC/DC biased [10]. A pulsed output is attainable for almost any of the carbon dioxide laser types; however, the ratio of peak to average power is relatively low due to the limited energy storage capability of the gaseous medium. A constant supply of gases is required for the carbon dioxide lasers, except for the sealed discharge tube models, which must be replenished periodically. The cost of the gas is minimal, except for helium, which can be expensive. Optics are made from Zinc Selenide (ZnSe), Gallium-Arsenide (GaAs) or from various salt crystals (NaCl or KCl). The salts possess poor hygroscopic properties and are limited mainly to research applications. Optical fiber technology has not yet been developed for beam delivery in gas lasers; however, flexible metal conduits with mirrored interior surfaces have been developed to carry up to 2 kW of laser power.

The *excimer laser* is an increasingly popular type of gaseous laser. The term "excimer" originates from "excited dimer," a compound of two identical species which exists only in an excited state [9]. Some excimer lasers use xenon (Xe_2) and fluorine (F_2) as the lasing material, but the majority of excimer lasers use an excited complex (exciplex) of a noble gas and halogen atom as the actual lasant. Typical excimer complexes include argon fluoride (ArF), krypton fluoride (KrF), xenon fluoride (XeF), and xenon chloride (XeCl). These compounds only exist temporarily when the noble gas is in an excited electronic state. The bond is very strong but lasts only a few nanoseconds. When the noble gas atom is no longer excited, each compound molecule dissociates into its elemental components; this dissociation is accompanied by a release of the binding energy in the form of a photon. The output wavelengths of the excimer lasers vary from 0.193 to 0.351 microns in the ultraviolet to near-ultraviolet spectra (Fig. 2.13).

ArF	.1933μm
KrF	.2484 - .2495μm
XeF	.3490 - .3540μm

XeCl .3080μm

These compounds can be formed by inducing the Noble gas (Ar, Kr, or Xe) of the compound into an excited state with an electrical discharge, an electron beam, or a combination of the two. In the electrical discharge method, an electrical discharge provides the initial energy for excitation of the noble gas [24]. The discharge also initiates an ionization process in the halogen atom, which releases energy to sustain the excitation of the noble gas electrons. The electron energy of the discharge is small (four to six electron-volts); consequently, the ionization process is the dominant mechanism for excitation. Through repetition of high power pulses, a set of electrical discharges is sufficient to maintain a high average power in the laser beam. In the method involving electron-beam pumping, the high energy of the electron beam (1-2 MeV) results in ionization of atoms. The ionized atoms attract neutral atoms to form ionized molecules which lose some energy and become excimers. The electron beam provides sufficient energy to sustain continuous excitation in the noble gas, so ionization of the halogen does not have a significant influence on excitation.

The excimer is usually formed in a rectangular resonator cavity. The output beam originates from the resonator and the discharge is transverse. The emission spectrum of photons is a wide (100-200Å) continuous band [24]. This width causes the stimulated cross section to be small, so that high excimer concentrations are required for amplification. Often, all the gases may be used with a single excimer laser; however, the laser must be conditioned between each gas. The halogens react with the materials of the laser, so an inert gas such as He or Ar needs to be flushed through to break up the compounds.

Excimer lasers produce high-powered, pulsed beams, with average power over 100W and average pulse repetition rate of 1000 pulses per second. The problem with a

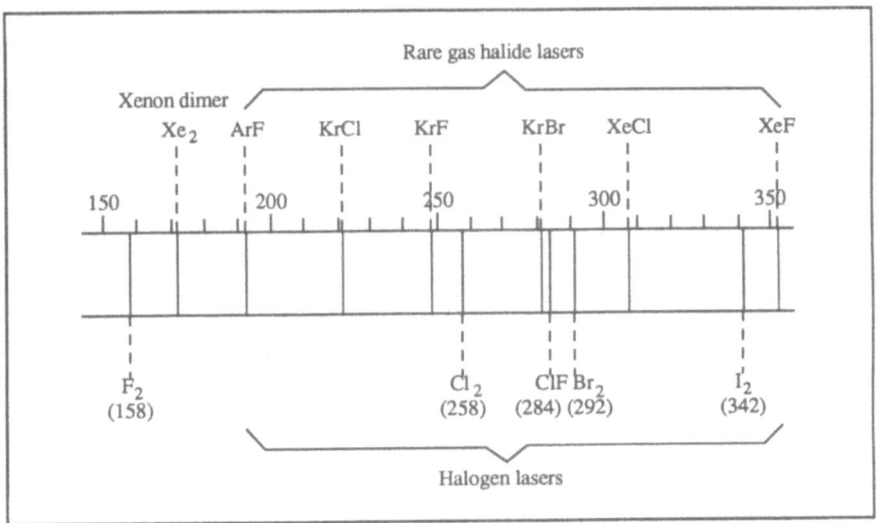

FIGURE 2.13   Spectra for Different Types of Excimer Lasers

high average power is that the lasing medium and the resonator cavity are both heated during the lasing process. At the same time, high power levels must be reached in order to achieve reliable periodic switching [9]. In the excimer compound, both intermediate products (F^-, Ar_2^+, etc.) and active states of the molecule (KrF^*, XeF^*, etc.) absorb the laser radiation with an absorption coefficient between 1×10^{-3} cm^{-1} and 5×10^{-3} cm^{-1}. This fact limits the length (L) of the laser to 2-3 m because the coefficient x L must be less than or equal to one; otherwise, all of the beam energy will be absorbed by the excimer and no laser beam can be produced. This shorter length limits the maximum output energy of the excimer laser, which makes the range of materials it can process much narrower than that for the CO_2 laser. The average power of an excimer laser pumped by an electrical discharge is limited by the instability of the discharge, the inhomogeneity of the excimer, and the chemical and thermal effects on the medium. In KrF, XeF, and XeCl lasers, this problem is addressed by driving the gas mixture rapidly and perpendicularly to the discharge zone.

Excimer lasers are used to machine solid polymer workpieces, remove polymer films from metal substrates, remove metal films from polymer substrates, micromachine ceramics and semiconductors, and mark thermally sensitive materials, among other things [9]. The characteristics of the excimer laser beam make it ideal for some material removal applications. The high energy of each photon from an excimer laser reduces the interaction time between the irradiated and bulk material of the workpiece; therefore, the heat affected zone is minimized. This is extremely desirable for surgical and polymer processing applications and can be applied to other materials such as ceramics as soon as higher average power models are developed. The use of excimer lasers for cutting, drilling and marking has increased since their development in the 1970's. Processing with excimer lasers results in higher precision and reduced heat damage zones compared with CO_2 and Nd:YAG lasers. The increased demand for excimer lasers in recent years has resulted in the development of more reliable lasers by manufacturers. The fact that industry has become more aware of the applications of excimer lasers has caused their use in materials processing to become more acceptable.

The physics of the material removal process for excimer lasers is different than that for CO_2 or Nd:YAG lasers. Instead of removing material through melting or vaporization, where the material is heated from solid to liquid and/or gaseous states, the excimer laser removes material through ablation, breaking the chemical bonds of the material until it dissociates into its chemical components. In ablation, no liquid or gaseous phases of the material are present. The ablation process occurs as the UV pulses of radiation from the excimer laser make contact with the workpiece surface. Organic materials absorb photons in a thin layer near the surface, which break the bonds in the organic compounds. When the rate of bond breaking exceeds a critical value, the material decomposes. In this manner, the pulse of the excimer beam removes a fraction of a micrometer-thick piece of material layer by layer.

Another difference between excimer and CO_2 or YAG lasers is that the latter focus the beam to a small area and traverse the workpiece. Excimer lasers produce large area beams that are masked through a template to achieve the desired cutting area (Fig. 2.14). After being passed through a mask, the beam is focused onto the workpiece through a lens. From optics:

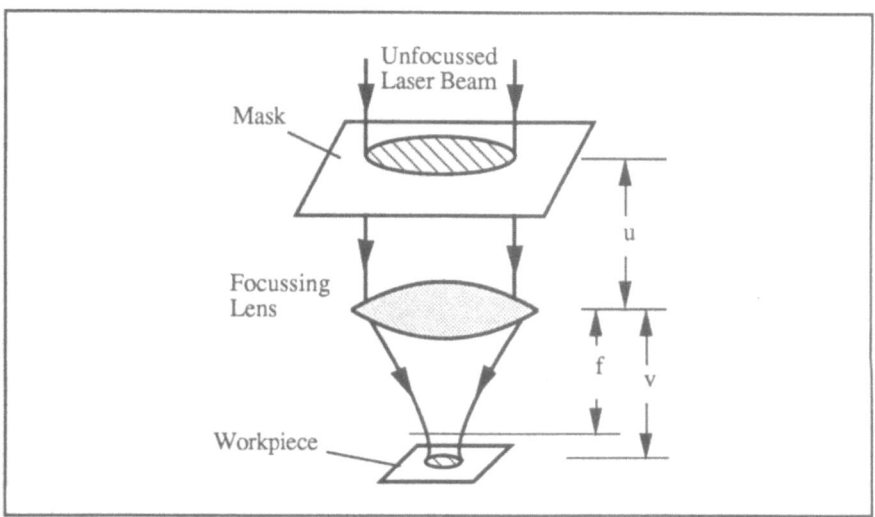

FIGURE 2.14   Configuration for Laser Machining With an Excimer Laser

$1/u + 1/v = 1/f$
u = distance from mask to lens
v = distance from lens to workpiece
f = focal length of lens

For values of u greater than 2 x f, the image on the workpiece will be inverted and demagnified, with a demagnification factor $d = |u/v|$. The energy density of the laser beam on the workpiece is proportional to the square of the demagnification:

$$ED_w = ED_0 d^2$$

(2-2)

where ED_w is the energy density at the workpiece and ED_0 is the energy density of the unfocussed beam. The energy density at the workpiece can be changed by adjusting the degree of demagnification. For example, high material removal rates for etching with an excimer laser can be achieved at $ED_w < 1$ J/cm^2 [9]. Since unfocussed excimer laser beams usually have energy densities from 100 to 200 mJ/cm^2, processing of polymers, for example, requires a demagnification factor of two to three. An added advantage is that the dimensional accuracy of the shape machined in the workpiece is high compared with the template shape, since there is no relative motion between the beam and the workpiece.

The mask is made of a metal that reflects most of the incident laser irradiation, so the material underneath the mask is not removed in the cutting process. Excimer lasers are excellent for machining repetitive patterns because the use of the mask allows for a series of holes or slits to be processed at the same time. This method is much more efficient than the use of a CO_2 or Nd:YAG laser, whose use requires that each hole or slit be cut individually. For example, an excimer laser can drill 5000 holes in a polyimide sheet in approximately three seconds, while the same process would require about 50 seconds with a CO_2 or Nd:YAG laser [9].

If the area of the workpiece to be processed exceeds 1cm x 1cm, or the patterns to be cut are not repetitive, contact masking is used, where the mask is held close to the surface of the workpiece while the laser beam passes over it. The material exposed to the beam through the mask openings is ablated. Another method for machining large areas is conformal masking, where a thin metal layer mask is deposited on the workpiece.

2.4.2   Liquid Lasers

Liquid lasers are primarily *dye lasers* which utilize large organic dye molecules as the lasing medium (Fig. 2.14) [19]. The dyes can absorb radiation from a wide range of frequencies and have hundreds of overlapping spectral lines from which they can lase. These lasers are designed so that the frequency at which they lase can be varied, and are called "tunable." The spectral ranges of the dyes encompass the visible spectrum and parts of the infrared and ultraviolet spectra. The dyes are normally circulated axially through a discharge tube. Continuous operation is difficult to achieve because population inversion is difficult to maintain; therefore, a pulsed mode of operation with an average power up to 10 mW is generally used. Excitation is usually provided traversely by another pulsed laser (typically Ar, N_2, or Nd:YAG), pulsed flashlamp, or focussed continuous laser. The tunability of these lasers makes them desirable for spectroscopic investigation or photochemical applications.

2.4.3   Solid State Lasers

Solid lasers use ions suspended in a crystalline matrix to produce laser light as shown in Figure 2.15. The ions or *dopants* provide the electrons for excitation, while the crystalline matrix propagates the energy between ions. The two main classes of dopants in the laser medium are Chromium (Cr^{3+}) for ruby lasers and Neodinium (Nd^{3+}) for *Nd:YAG* and *Nd:glass* lasers [20]. Ruby lasers generally are no longer used because of low energy efficiency and low achievable beam power. The Nd:YAG and Nd:glass lasers are in general very similar to each other in structure and lasing action. Excitation is achieved by krypton or xenon flash lamps, and an output wavelength of 1.06μm in the near infrared region of the spectrum is obtained. The Nd:glass laser uses a glass host material for the neodymium ions. The glass rods have the advantage that they can be grown to larger sizes more economically than the YAG crystals, but glass has a lower thermal conductivity which limits the pulsed operation of the Nd:glass laser. Therefore, Nd:glass lasers are used in applications which require low pulse repetition rates and high pulse energies (up to 100 Joules per pulse) [20]. In general, the pulse operation of the neodymium lasers make them desirable for hole piercing and deep keyhole welding applications.

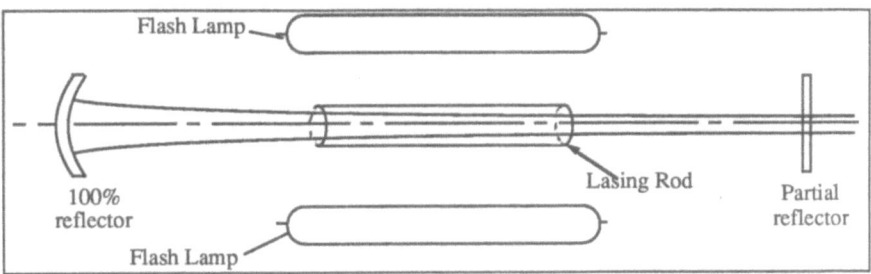

FIGURE 2.15   Schematic of a Solid Laser

The host material in Nd:YAG lasers (Fig. 2.16) is a complex crystal of $\underline{Y}$ttrium-$\underline{A}$luminum-$\underline{G}$arnet (YAG) with the chemical composition $Y_3Al_5O_{12}$. The YAG crystal has a relatively high thermal conductivity which improves thermal dissipation in the laser cavity, so continuous wave operation up to a few hundred Watts is possible [15]. When operated in a pulsed mode, high pulsing rates can be achieved, and average powers of up to 1 kW are available. The actual lasant is the Nd^{+3} ions which have been doped into a YAG crystal. The YAG crystal is transparent and colorless. When doped with approximately 1% Nd, the crystal takes on a light blue color.

Solid-state lasers usually use krypton or xenon flashlamps for optical pumping. Krypton flashlamps have low operating current and high energy efficiency, so they are useful for continuous wave operation. Xenon flashlamps have a better spectral coupling with the energy states of the neodymium ions than their krypton counterparts; therefore, they can support the higher current density discharges required for pulsed mode operations. The optical pumping cavity is most commonly shaped as an ellipse, where the lasing rod and the flashlamp are each placed at one of the foci of the ellipse. The distance between the lasing rod and the flashlamp is minimized so that the direct coupling between the two is maximized. In addition, any light radiated by the flashlamp that is not directed at the lasing rod is reflected off of the walls of the elliptical cavity at the lasing rod. This combination of direct and indirect pumping results in low electrical power requirements and an acceptable energy efficiency for both continuous wave and pulsed operation.

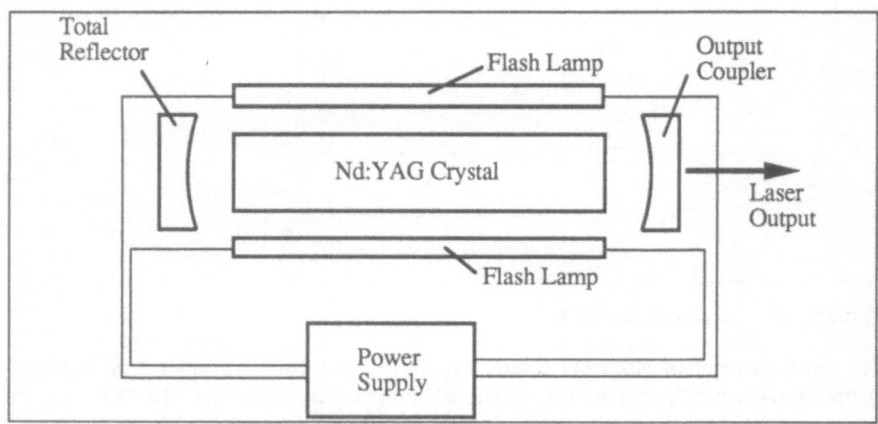

FIGURE 2.16   Schematic of a Nd:YAG Laser

Generally, a cooling system is required for operation of Nd:YAG lasers. With an efficiency of about 3%, a typical Nd:YAG produces thirty times as much waste heat as laser output; this heat must be removed in order to ensure proper laser operation. Flooding the optical compartment with water can remove waste heat; however, the water absorbs a significant amount of the flashlamp energy, and the turbulence of the water can produce optical distortion and imaging problems. These problems can be overcome by flowing water over the outside of the optical cavity and by encasing the lasing rod and flashlamp with transparent cooling jackets. This arrangement minimizes the amount of water that the flash light must travel through and allows better control of the water flowing in the optically distorting areas between the flashlamp and the lasing rod. Furthermore, the flashlamp electrode assemblies should be cooled to prevent thermal loading, and deionized water should be used to prevent any electrical conduction through the water.

The output characteristics of a Nd:YAG laser can be altered by varying the pumping discharge waveform. For high pulse energy in a pulse time of one millisecond, a long lamp pulse is used to obtain a six to seven Joule pulse composed of many spiked oscillations within the pulse. Such a laser can operate at several hundred cycles per second and is useful in laser cutting applications. Laser beam pulse frequency and shape can be tailored by using *Q switching*, where a shutter moves rapidly in and out of the path of the beam. In this manner, beam output is interrupted until a high level of population inversion and energy storage is achieved in the resonator. If the optical cavity is switched from no reflections (low Q) to near total reflection (high Q), the cycle can be optimized to build up the maximum population inversion before the pulse is generated (switched from low to high Q). This results in a beam pulse with a high energy (up to one joule) and a short pulse period (down to 10 ns), which is especially useful for deep hole drilling applications.

Continuous wave laser operation using a continuous flashlamp can be applied to laser cutting. Some cutting and marking operations require a higher peak power, so a pulsed beam output using Q switching is sometimes used. Average output power is comparable to similar continuous outputs. Pulse durations from 10-200 ns are used with pulse frequencies up to 5 kHz, resulting in peak power levels which are a few orders of magnitude greater than average beam power.

The operating cost for a Nd:YAG laser includes the replacement of the flashlamps, which cost $100-150 and last from 50,000 to 200,000 hours [20]. Optics suitable for transmitting light at a wavelength of 1.06 μm are made from fused quartz, which is a commonly used optical material and is relatively rugged compared to the transmissive optical materials required for carbon dioxide laser emissions. Also, fiber optic materials for beam delivery with good hygroscopic properties and flexibility have been developed for Nd:YAG laser radiation.

2.5 Laser Equipment Characteristics

After the use of laser machining has been decided upon, a suitable laser must be selected. Since industrial laser systems have high capital and operating costs, the choice of a particular laser must be carefully matched with both the machining application and the materials to be processed in order for the operation to be economically viable. The five major characteristics of the laser beam which are critical in determining the type of laser to use for a particular application are *the beam's power, wavelength, temporal mode, spatial mode,* and *focal spot size.* Other practical and economic considerations include the output stability of the power and TEM mode, the physical size of the unit, projected lifetime of the laser, gas or flash lamp consumption, availability of service and parts, the power requirements, and the efficiency of the laser. These secondary characteristics vary widely depending on the make and model of the laser. Detailed information can be found in the service manual for the specific laser.

2.5.1 Laser Power

Output power is the most basic characteristic of a laser. Purchase of an under-powered laser system will result in increased processing time or an inability to perform machining operations on the desired materials. Capital expenses for a laser system increase with laser power, and purchase of an over-powered laser system may result in excessive expenses. In general, the highest continuous wave power is obtained from CO_2 lasers, while Nd:YAG lasers provide the highest peak power for pulsed operation. The amount of laser power required is determined by examining the optical and thermal properties of the workpiece material or group of materials to be machined. The *thermal* properties can be divided into two basic categories: *fixed* and *loss properties.* The first category determines the amount of energy required to melt and vaporize the material. This includes heat capacity, latent heat, and heat of vaporization. For example, ceramic materials generally require higher laser power to machine than plastics due to their higher latent heats. *Loss properties* are important because they determine the energy transmitted to the surrounding material during processing. Thermal diffusivity is most important in transient operations, while thermal conductivity is more important in steady-state applications.

The laser's optical properties affect the surface of the workpiece where the laser beam impinges. Of the optical material properties, the *absorptivity of the material* has the largest influence on laser power requirements. The material absorptivity determines the fraction of the impinging radiation energy that is actually absorbed by the material. The remainder of the beam energy is reflected back into the environment. Thus, laser power must be adjusted so that the required amount of power for achieving the desired material removal at the desired processing rate is absorbed by the surface, instead of only impinging on the surface. The absorptivity value can vary depending on the wavelength of the beam impinging on the surface, surface roughness, temperature, phase of the material, and the use of surface coatings. The process rate also is related to the required laser power; however, this relationship is complex, since the process rate will also affect the energy losses of the process and the power will affect the material/laser coupling. Estimation of power requirements can be performed by using models of the specific laser machining process to relate the material properties, operating parameters such as laser power and process rate, and material removal characteristics such as depth of cut, hole depth, and the shape of the cut.

2.5.2 Wavelength

The *wavelength* is the characteristic spatial length associated with one cycle of vibration for a photon in the laser beam. The optical resonator of the laser must be properly designed to produce the correct wavelength; the length of the resonator must be such that the laser beam traverses an integral number of half wavelengths of the desired frequency. Any variations in this distance due to thermal expansion or mechanical vibrations may cause the laser to vary its output wavelength, mode or efficiency. Thermal expansion of the resonator components can be controlled by water cooling. Large mechanical structures are used to provide a stiff base for dampening mechanical vibrations to the resonator, particularly to the mirror components. The selection of optical accessories such as lenses, mirrors, polarizers, and windows also depend upon the wavelength of the laser. If fiber optics are intended for beam delivery, suitable optical fiber materials are available for only a few specific laser wavelengths.

The absorptivity of materials can be highly dependent on the wavelength of incident light, and thus certain lasers will be more suitable for the processing of different classes of materials. For example, some metals such as aluminum and copper show low absorptivity at a wavelength of 10.6μm, which is the characteristic wavelength for CO_2 lasers. Therefore, in order to machine these materials effectively, either a high power laser or one with a different wavelength must be used. Materials can be machined more effectively by using a Nd:YAG laser with a 1.06μm wavelength, whereby the copper and aluminum absorptivity values are much higher.

2.5.3   Temporal Mode

Lasers can operate in either a continuous wave (CW) mode or a pulsed beam mode (Fig. 2.17). In continuous wave mode the laser beam is emitted without interruption. In the pulsed beam mode, the laser beam is emitted periodically. Pumped energy is stored until a threshold energy is reached, and then the stored energy is discharged rapidly into the laser cavity. Through this process, short duration pulses with high energy densities can be generated from lower levels of continuous electrical power. CW operation offers the advantage of smooth surfaces after machining. However, the surface quality improvement is offset by the high electrical power input required to maintain a continuous beam. Pulsed beam operation allows deeper drilling or cutting depths to be achieved compared to a continuous beam operating at a given beam power, but pulsed operation may result in larger surface irregularities in the machined parts due to a periodic beam output.

The temporal mode capability of a laser depends greatly on the lasing medium. Typically, solid lasers operate best in pulsed mode and at relatively lower powers in continuous mode. The solid lasant provides a large population of excited ions because of the relatively high density of the material; however, there is a limited ability to regenerate these excited ions due to the amount of time needed to return them to the ground state. Thus, high energy pulses can be generated by the rapid stimulation of the entire lasing medium. Continuous output is limited by the inability to quickly return ions to the ground state for re-excitement. Also, there are thermal limits to the amount of heat the lasant can absorb and the rate the heat can be removed; these limits affect pulse repetition rates and continuous power ratings.

Gas lasers typically operate in a continuous mode with limited pulse capability. The gas lasant does not contain as many excitation sites as a solid lasing material, and therefore, the pulse energy is limited. Typically, the continuous power ratings are also limited by the low density of the laser medium for lasers in which the gas in the laser cavity is

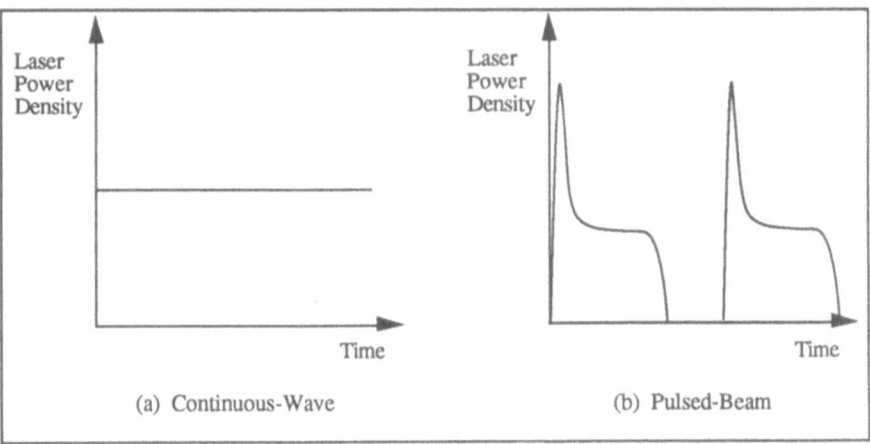

(a) Continuous-Wave   (b) Pulsed-Beam

FIGURE 2.17   Laser Beam Temporal Modes

not replenished. This can be increased many orders of magnitude by continuously flowing new lasing material into the optical resonator. The incoming gas is in the ground state and is always ready for excitation and stimulation. Also, gases which have undergone stimulation and need to return to the ground state are removed from the laser cavity. The flow of gas effectively reduces the cycle time of the emission process and greatly increases the power potential of the laser. This technique also improves the pulsing capability, as do altering of the lasing gas mixture and using specialized electrical pumping techniques.

The selection of a laser and an operating mode will depend strongly on the desired machining operation. Pulsed operation is usually best for deep penetration processes. The concentration of energy in each pulse leads to a small percentage of energy lost through conduction into the workpiece or dissipation to the environment. In addition, the coupling of the material and the laser may require the use of higher power densities available only by pulsing. For the same reason, pulsing is used to minimize the heat affected zone in materials that are sensitive to elevated temperatures, such as polymers. Continuous power operation is used when high average power is required, which is important for achieving high material removal rates. When material/laser coupling is good, continuous power operations will not suffer from large losses. Also, surface finish is often an important concern. Continuous operation provides smooth cutting surfaces, while pulsing can create a wavy surface due to its periodic nature.

2.5.4   Spatial Mode

As mentioned above, resonator design is crucial to the production of the proper wavelength of light. Similarly, the phase of the electromagnetic wave may differ by resonator design, resulting in changes in the laser beam spatial profile. The beam profile can be characterized by its *Transverse Electromagnetic Mode* (TEM) [3]. TEM modes are normally written in the form of TEM_{nm}. The subscripts n and m denote the number of nodes in directions orthogonal to the beam propagation, such as TEM_{00} or TEM_{01} (Fig. 2.18). TEM_{00} has a Gaussian spatial distribution and is usually considered the best mode for laser machining because the phase front is uniform and there is a smooth drop off of irradiance from the beam center. This minimizes diffraction effects during focussing and allows the generation of small spot sizes. Correct resonator design can minimize non-Gaussian modes. Higher order modes have more rapid diffraction characteristics. In a resonator designed to produce a beam with a TEM_{00} spatial mode, a large part of the higher order modes is diffracted beyond the reflection mirrors and discarded. In order to minimize the occurrence of higher-order modes in the laser beam, an aperture can be used at the midpoint of the resonator. The aperture is designed so that it is large enough to pass the Gaussian components, but higher order modes are attenuated because they diverge after hitting the reflecting mirrors and cannot pass through the narrow diameter of the aperture. In some applications such as welding or heat treating, use of higher-order modes is desirable because these applications require uniform heating over the surface area more than localized high energy densities. By using a higher-order mode beam, energy can be distributed evenly over a larger spot size.

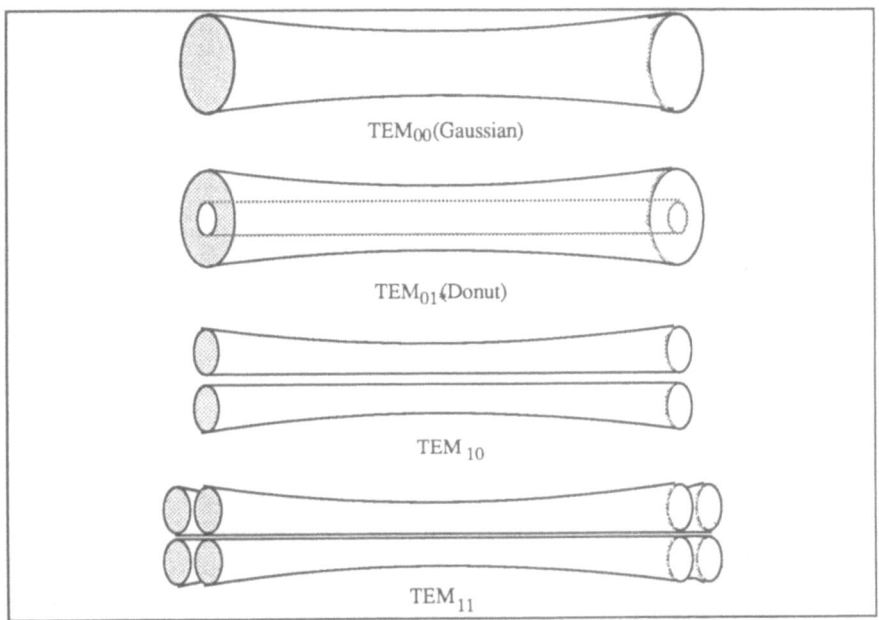

TEM_{00}(Gaussian)

TEM_{01}(Donut)

TEM_{10}

TEM_{11}

FIGURE 2.18   TEM Modes

2.5.5   Focal Spot Size

In materials processing, *irradiance (power per unit area)* of the laser beam at the material surface is of prime importance [15]. Irradiances great enough to melt or vaporize any material can be generated by focussing a laser beam. The maximum irradiance is obtained at the focal point of a lens, where the beam is at its smallest diameter (Fig. 2.19); the location of this minimum diameter is called the *focal spot*. Irradiance values of billions of watts per square centimeter can be obtained at the focal spot. The imperfections of the optical components and diffraction effects limit the size of the obtainable focal spots.

Several factors influence focal spot size. First, focal spot size is directly related to the quality of the incoming beam, which can be quantified by the divergence of the beam (Fig. 2.19). A laser beam with a small divergence can be focussed to a smaller spot than a beam with high divergence [3]. Second, focal spot size is influenced by diffraction. When focussing a diffraction-limited laser beam with a lens, a longer focal length or higher f-number corresponds to a larger focussed spot diameter. Finally, the diameter of the incoming laser beam affects the focal spot size. When restricted to specific optics, the only method of producing smaller focal spot sizes is to *increase* the incoming laser beam diameter.

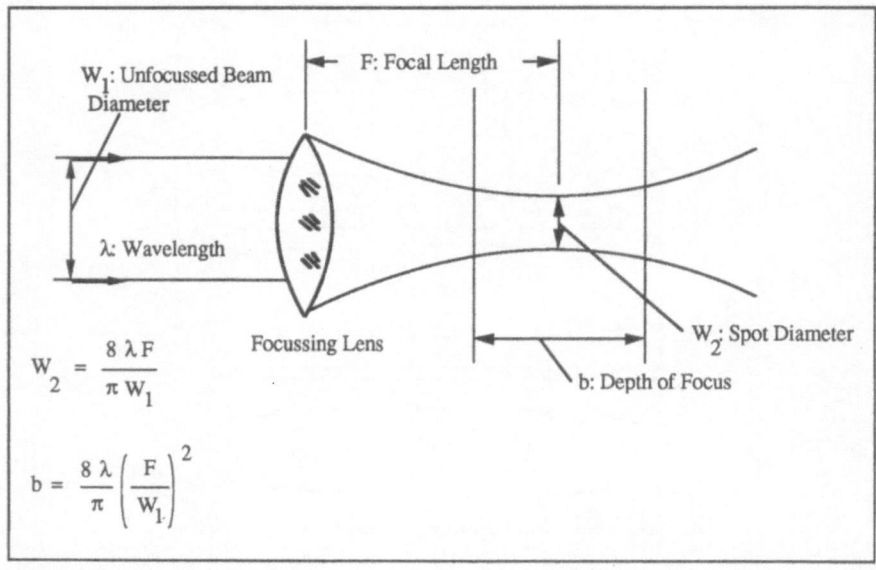

FIGURE 2.19   Calculation of Focal Spot Diameter and Depth of Focus

References

1. Belforte, D., and M. Levitt (eds.), *The Industrial Laser Annual Handbook*, Pennwell Pub., Tulsa, Oklahoma, 1987.

2. Berloffa, E., "Laser Cutting–Principal System Aspects and Practical Results," *Proceedings of the SPIE – High Power Lasers and Their Industrial Applications*, Vol. 650 (1986), 289-294.

3. Coherent, Inc. Staff, *Lasers: Operation, Equipment, Application, and Design* McGraw-Hill, New York, 1980.

4. Courtney, J., D. Wheatley, P. Oakley, and R. Crager, "Advanced Modular Fast Axial Flow Laser," *Proceedings of the First International Conference on Lasers in Manufacturing*, IFS Publications, Ltd., Amsterdam, 1983, 235-254.

5. Duley, W.W., *Laser Processing and Analysis of Materials*, Plenum Press, New York, 1983.

6. Fantini, V., B. Incerti, and V. Filzi, "High Performance 2.5 kW Industrial CO_2 Laser," *Proceedings of the SPIE – High Power Lasers and Their Industrial Applications*, Vol. 650 (1986), 36-51.

7. Feinberg, G., "Light," *Scientific American* , Vol. 219, No. 3 (Sept. 1968), 50-59.

8. Hugel, H., "Rf-Excitation of High Power CO_2 Lasers," *Proceedings of the SPIE – High Power Lasers and Their Industrial Applications* , Vol. 650 (1986), 2-9.

9. Humphries, M., H. Kahlert, and K. Pippert, "The Excimer Laser on its Way to Industrial Application," *Proceedings of the First Conference on Lasers in Manufacturing,* IFS Publications, Ltd., Amsterdam, 1983, 255-262.

10. Laos, O., "Evaluating a CO_2 Industrial Laser System," *Proceedings of the First Conference on Lasers in Manufacturing*, IFS Publications, Ltd., Amsterdam, 1983, 21-30.

11. Maisenhalder, F., "High Power CO-Lasers," *Proceedings of the SPIE – High Power Lasers and Their Industrial Applications*, Vol. 650 (1986), 85-91.

12. Ogorek, M., and G. Farnum, "Lasers: The Light Touch for Industry," *Manufacturing Engineer* (September 1985), 71-78.

13.   Parker, D., *Industrial Lasers and Their Applications*, Prentice-Hall, Englewood Cliffs, N.J., 1985.

14.   Patel, C., "High-Power Carbon Dioxide Lasers," *Scientific American,* Vol. 219, No. 3 (September 1968), 23-33.

15.   Ready, J., "Lasers–Their Unusual Properties and Their Influence on Applications," *Lasers in Modern Industry,* Society of Manufacturing Engineers Marketing Services Dept., Dearborn, MI, 1979, 17-38.

16.   Schaffer, G., "Lasers in Metalworking," *Lasers in Modern Industry,* Society of Manufacturing Engineers, Dearborn, MI, 1979, 3-16.

17.   Schawlow, A., "Laser Light," *Scientific American,* Vol. 219, No. 3 (September 1968), 120-136.

18.   Spalding, I., "High Power Lasers for Processing of Materials - A Comparison of Available Systems," *Lasers in Modern Industry,* Society of Manufacturing Engineers, Dearborn, MI, 1979, 39-47.

19.   Spalding, I., "Which Wavelength? – How to Select a Suitable Laser," *Proceedings of the First International Conference on Lasers in Manufacturing,* IFS Publications, Ltd., Amsterdam, 1984, 229-234.

20.   Weber, H., "High Power Nd-Lasers for Industrial Applications," *Proceedings of the SPIE – High Power Lasers and Their Industrial Application,* Vol. 650 (1986), 92-100.

21.   Wollermann-Windgasse, R., F. Ackermann, J. Wick, and W. Brix, "Rf-Excited High Power CO_2-Lasers for Industrial Material Processing," *Proceedings of the SPIE – High Power Lasers and Their Industrial Applications,* Vol. 650 (1986), 29-35.

22.   Luxon, J., D. Parker, and P. Plotkowski, *Lasers in Manufacturing: An Introduction to the Technology,* IFS Publication, New York, 1987.

23.   Eloy, J., *Power Lasers,* John Wiley, New York, 1985.

24.   Kompa, K., "Important Parameters of Excimer Lasers: Relevant Topics of Current Research," *Proceedings of the SPIE – High Power Lasers and Their Industrial Application,* Vol. 650 (1986), 75-81.

3
Basics of Laser Machining

This chapter introduces the basic physical mechanisms in laser machining processes. Laser machining can be divided into one, two and three-dimensional processes by differentiating the kinematics of the erosion front during beam/material interaction. All laser machining processes exhibit common characteristics such as molten layer formation, possible plasma formation, and beam reflection from the erosion front. This chapter also discusses the major components required for a laser machining system: the laser, the beam delivery subsystem, the workpiece positioning subsystem, and auxiliary devices. Finally, possible real-time sensing techniques and approaches for closed-loop control of laser machining processes are discussed.

3.1 Laser Machining as a One, Two, and Three-Dimensional Process

Laser machining can replace mechanical material removal methods in many industrial applications, particularly in the processing of difficult-to-machine materials such as hardened metals, ceramics, and composites. Furthermore, laser beams themselves make new material removal methods possible, due to their unique characteristics.

- *Laser machining is a thermal process.* The effectiveness of laser machining depends upon the thermal properties and, to a certain extent, the optical properties rather than the mechanical properties of the material to be machined. Therefore, materials which exhibit a high degree of brittleness or hardness and have favorable thermal properties such as low thermal diffusivity and conductivity are particularly well-suited for laser machining.

- *Laser machining is a non-contact process.* Since energy transfer between the laser and the material occurs through irradiation, no cutting forces are generated by the laser, leading to the absence of mechanically-induced material damage, tool wear and machine vibration. Moreover, the material removal rate for laser machining is not limited by constraints such as maximum tool force, built-up edge formation, or tool chatter.

- *Laser machining is a flexible process.* When combined with a multi-axis workpiece positioning system or robot, the laser beam can be used for drilling, cutting, grooving, welding, and heat treating processes on a single machine. This flexibility eliminates the transportation necessary for processing parts with a set of specialized machines. Also, laser machining can result in higher precision and smaller kerf widths or hole diameters than comparable mechanical techniques.

However, because laser machining is a primarily thermal process, it has a few disadvantages:

- *Low energy efficiency.* In most laser machining techniques, the removal of material occurs by melting or vaporizing the entire volume to be removed. Since this phase change of the material occurs on an atom-by-atom basis, laser machining requires significantly higher energy inputs and processing times than equivalent mechanical processes. Further development of industrial lasers may increase the efficiency of converting electrical energy to beam energy. Also, the development of three-dimensional laser machining techniques provides an energy-efficient material removal method with a high degree of flexibility.

- *Material damage.* During laser machining, high power densities are introduced on the surface of the workpiece in order to raise the temperature of the volume to be removed to the melting or vaporization point. In metals, the conduction heat resulting from the high energy density on the workpiece creates a heat-affected zone in the vicinity of the erosion front. In plastic and polymer-matrix composites, material decomposition may occur as a result of elevated temperatures, which cause the breakdown of the polymer into charred residues and gaseous products.

In general, laser machining can be divided into *one-, two-,* and *three-dimensional* processes (Fig. 3.1). Differentiation between these categories can be made by examining the shape and the kinematics of the erosion front, namely the region in the workpiece where material removal takes place. Since the laser beam is a directional heat source, it can be viewed as a one-dimensional line source with a line thickness equal to the beam diameter.

   In the case of a one-dimensional process (drilling), the laser beam is stationary relative to the workpiece. The erosion front, located at the bottom of the drilled hole, propagates in the direction of the line source in order to remove material. In the case of a two-dimensional process (cutting), the laser beam is in relative motion with respect to the workpiece. Material removal occurs by moving the line source in a direction perpendicular to the line direction, thereby forming a two-dimensional surface. The erosion front is located at the leading edge of the line source. For three-dimensional machining, two or more laser beams are used, and each beam forms a surface through relative motion with the workpiece. The erosion front for each surface is found at the leading edge of each laser beam. When the surfaces intersect, the three-dimensional volume bounded by the surfaces is removed.

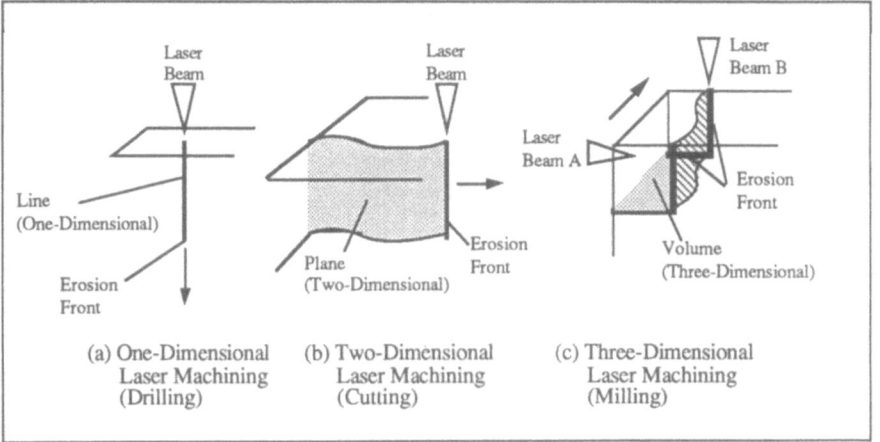

FIGURE 3.1   Schematic of One, Two and Three-Dimensional Laser Machining

Three major issues in any laser machining process are:

- Material removal rate
- Dimensional accuracy
- Surface quality

The material removal rate is governed in each case by the propagation speed of the erosion front. In laser drilling (a one-dimensional process), material removal rate is determined by the speed that the erosion front moves in the beam direction. In laser cutting (a two-dimensional process), the scanning velocity determines the rate at which the two-dimensional surface increases in the workpiece. In three-dimensional laser machining, two-dimensional surfaces produced by two laser beams define a three-dimensional volume of material to be removed. The speed at which these two surfaces propagate determines the time required to remove a given volume of material.

Dimensional accuracy is determined particularly by the hole taper for laser drilling, kerf geometry for laser cutting, and groove shape for three-dimensional machining. Surface quality for all laser machining processes is related to factors such as surface roughness, dross formation, and heat-affected zone.

3.1.1   Drilling (One-Dimensional Laser Machining)

Laser drilling involves a stationary laser beam which uses its high power density to melt or vaporize material from the workpiece. This method is sometimes called *percussion* or *on center* drilling. In principle, laser drilling is governed by an energy balance (Fig. 3.2) between the irradiating energy from the laser beam and the conduction heat into the workpiece, the energy losses to the environment, and the energy required for phase change in the workpiece.

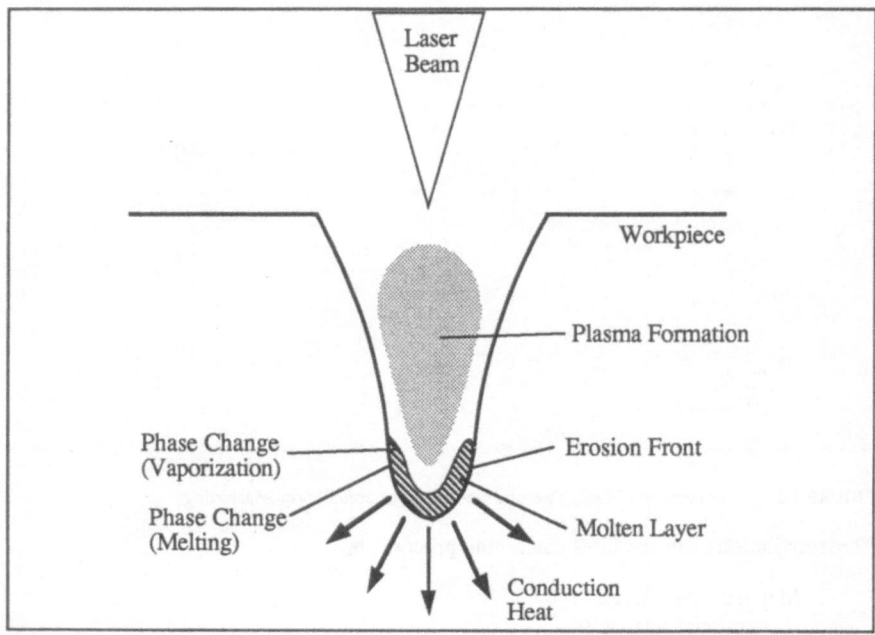

FIGURE 3.2   Laser Drilling

The incident beam energy has a spatial intensity distribution which in laser drilling is usually a Gaussian distribution produced by a laser operating in a TEM_{00} mode. The focussed beam radius is usually specified as the distance between the beam center and a point where the intensity is reduced from its maximum value at the beam center by a factor of e^2 (Fig. 3.3); the average diameter of the drilled hole may be less than the beam diameter due to various heat loss effects. These heat losses, primarily conduction to the interior of the workpiece and losses to the environment, divert beam energy away from the actual hole drilling process.

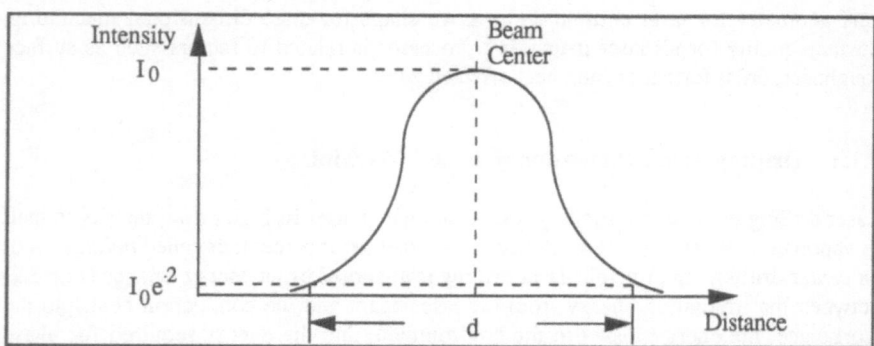

FIGURE 3.3   Spatial Intensity Distribution for TEM_{00} Laser Beam

Conduction heat, which occurs as a result of the temperature difference between the hole surface and the workpiece interior, is dependent upon the material thermal diffusivity, α, and the interaction time, t_i, which both define the thermal penetration depth:

$$\delta = \sqrt{\alpha \, t_i} \qquad\qquad (3\text{-}1)$$

A more detailed discussion of the characteristics of the temperature distribution and conduction heat inside the workpiece can be found in Chapter 5.

Energy losses occur due to a number of physical phenomena during laser machining:

- If the material removal process involves melting, the molten material may accumulate along the side and bottom of the hole, causing laser beam energy to be lost in two ways. First, energy may be expended to superheat the accumulated molten material in the hole above the melting point. Second, in percussion drilling, where a sequence of pulses is used to drill a hole, the molten material may resolidify between successive pulses; therefore, a portion of the beam energy during each pulse is expended to remelt the resolidified material.

- Plasma formation may occur when material is vaporized. An opaque cloud of vaporized material often forms above the interaction zone. This cloud partially absorbs the incoming beam energy and increases its temperature until a plasma is formed. In some cases, the heated plasma acts as a secondary heat source which improves the drilling process. However, the directionality of the plasma is difficult to control, and this causes dimensional accuracy problems. Use of an inert assist gas can help reduce plasma formation by removing vaporized debris from the path of the laser beam.

- The absorption of the laser beam energy depends on both the wavelength of the laser radiation and the spectral absorptivity characteristics of the material processed. Some metals, such as aluminum and copper, exhibit high reflectivity for CO_2 laser radiation (10.6 μm wavelength), so Nd:YAG lasers are more effective. Also, for metals and ceramics, the presence of a molten layer changes the absorptivity value. Furthermore, the absorptivity of a surface depends on its orientation with respect to beam direction. Absorption of beam energy is shown to reach a maximum value for angles of incidence above 80°. The beam energy not absorbed by the workpiece is reflected in a different direction from the incoming energy. For deep holes, multiple beam reflections may occur along the wall of the hole, thereby decreasing the availability of beam energy for material removal.

- The use of a gas jet during laser drilling may aid in cooling the erosion front through convective heat transfer. In situations where a high-pressure gas jet is used in tandem with a laser beam, a supersonic gas flow is formed and the thermal dissipation to the jet may become significant. With an increase in the thermal dissipation, more beam energy is required to maintain the melting/vaporization temperature at the erosion front.

Laser drilling has several advantages over mechanical methods:

- Due to the thermal nature of the laser drilling process, holes can be made on materials which are difficult to machine with conventional methods, such as ceramics, hardened metals and composites.

- Higher accuracies and smaller dimensions can be achieved with laser drilling than with conventional drilling methods. Depending on the focussing lens used, hole diameters between 0.001 in (0.018 mm) and 0.050 in (1.3 mm) are achievable. With the proper selection of values for beam power, beam pulse characteristics, focussing lens, and interaction time, the desired hole geometry can be obtained.

- High drilling rates can be achieved in a production environment by using a pulsed beam source. By coordinating workpiece motions with the pulse period, drilling rates above 100 holes per second can be achieved. The controlling of process variables allows rapid changes in hole diameters and hole shapes to be made in-process, eliminating the need for tool changes.

- The laser allows holes to be drilled at high angles of incidence to the surface (up to 80°). Shallow angle drilling is difficult to achieve mechanically due to tool deflections.

There are limitations to laser drilling in some cases, however:

- Holes with stepped diameters cannot be drilled using a laser.

- Due to instabilities in the laser drilling process, depth control in blind hole drilling is difficult. However, continuous monitoring of beam mode and beam power regulation can provide substantial benefits when incorporated into the laser drilling system in the form of a controller.

- For deep holes, the effects of beam divergence may become unacceptable. This can be compensated for by using a longer focal length lens or by continuously moving the focal point from the workpiece surface to a point at the workpiece interior.

There are several variations on laser drilling. The material removal rate for drilling of metals can be substantially increased by introducing a reactive gas to enhance the laser cutting process. In this case, chemical reactions between the workpiece material and the gas become an important secondary material removal mechanism. A drawback to this process is that while the laser beam is a directional heat source, chemical reactions tend to propagate in all directions and dimensional control of the reaction process is poor. Large-diameter holes (diameters above 0.050 in or 1.3 mm) can be produced by a trepanning method, in which the beam is scanned in a circular trajectory to obtain the final geometry. The trepanning method is actually a circular through-cutting technique, with the machining speed determined by the scanning velocity of the beam.

3.1.2 Cutting (Two-Dimensional Laser Machining)

In the laser through-cutting process, a kerf is created through relative motion between the laser beam and the workpiece surface. This process allows intricate two-dimensional shapes to be cut on a flat workpiece. The physical mechanisms for material removal and energy losses (Fig. 3.4) are similar to those for drilling, where the incoming laser beam energy is balanced by the conduction heat, energy for melting or vaporization of material, and heat losses to the environment. However, due to the relative beam/workpiece movement, the erosion front formed in front of the laser beam and the temperature field in the workpiece are stationary with respect to a coordinate system moving with the laser beam; therefore, laser cutting can be considered a steady-state thermal process. Since the workpiece thickness is equal to the depth of cut, conduction heat occurs in the plane of the workpiece. The temperature inside the workpiece is dependent on the distance to the erosion front and is independent of time.

When material is removed through melting, a molten layer forms at the erosion front. The accumulated molten material can be expelled out from the bottom of the kerf with the aid of a coaxial gas jet.

Because of the unique characteristics of the laser beam, laser cutting has several advantages:

- For most industrial materials with workpiece thicknesses up to 10mm, laser cutting produces a significantly higher material removal rate than mechanical cutting or shearing.

- Laser cutting produces kerf widths which are narrower than those achievable with mechanical cutting. This results in less material wasted during cutting operations.

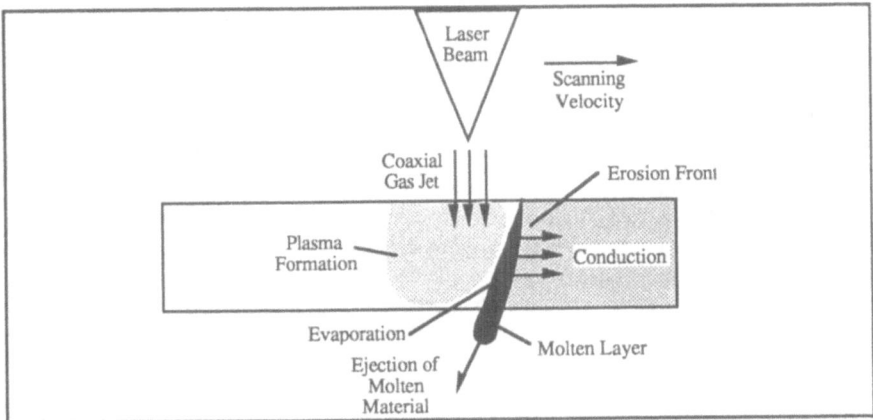

FIGURE 3.4   Laser Through-Cutting

- When coupled with a multi-axis position control system for the workpiece or beam, shapes can also be cut from curved workpieces. Conventional mechanical methods can only cut flat workpieces effectively. Lasers can be applied to trimming operations to remove flash and burrs from curved parts.

- For cutting of fibrous material such as wood, paper, or composites, the laser beam vaporizes the volume of material to be removed, thereby eliminating the residue and debris which remain after mechanical cutting. This reduces the need for solid waste collection and disposal and reduces the health hazard in the work environment.

The drawbacks of laser cutting in comparison to conventional methods are:

- Laser cutting effectiveness reduces as the workpiece thickness increases. Workpieces greater than 15mm in thickness generally cannot be cut effectively by modern industrial lasers.

- Laser cutting produces a tapered kerf shape, compared to the straight vertical kerf walls achievable by conventional methods. The kerf taper is a result of the divergence of the laser beam and becomes more pronounced as the workpiece thickness increases. The kerf taper can be reduced by adjusting the focal point of the laser beam to the interior of the workpiece instead of on the workpiece surface.

3.1.3 Turning/Milling (Three-Dimensional Laser Machining)

To make laser machining more applicable to bulk material removal, a concept for three-dimensional material removal has been developed. This method uses two intersecting laser beams to remove a volume of material. Unlike laser through-cutting techniques, each beam creates a blind groove in the workpiece through single or multiple passes. A volume of material is removed when the two grooves intersect. Turning operations can be accomplished by ring removal or helix removal (Fig. 3.5). The ring removal method uses two perpendicular beams to remove concentric rings from a workpiece. The helical removal method uses two angular beams to create a continuous thread. A three-dimensional volume of material is removed by creating two intersecting kerfs. Energy is only expended to melt and/or vaporize the material in two grooves, and the other removed material, or "chip," requires no energy expenditure. Therefore, the volume of material removed per unit energy expended by these 3-D processes is significantly greater than that for single-beam processes. The material removal rate can also be increased by chip formation to a level comparable to mechanical machining speeds.

For the case of laser milling, two laser beams are positioned at oblique angles from the workpiece surface to create converging grooves in a workpiece. The volume of material removed is prismatic in shape with a triangular cross section. By using small incidence angles between the laser beam direction and perpendicular direction to the surface, large material volumes can be removed, but the resulting part has a rough

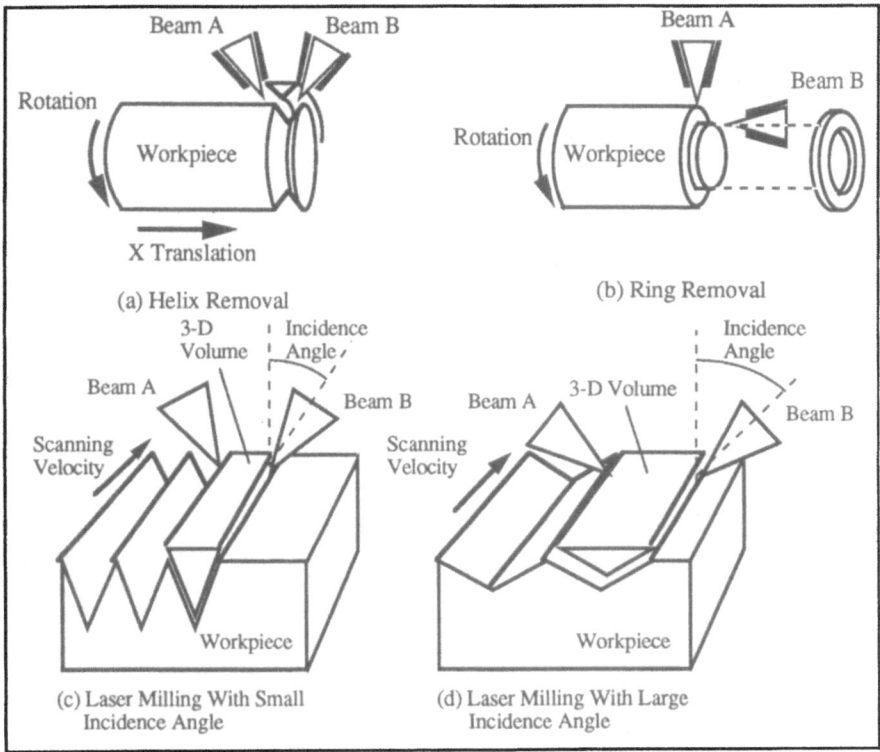

FIGURE 3.5   Three-Dimensional Laser Machining

surface finish. For large incidence angles, the material removal rate is reduced, but surface quality improves.

Since three-dimensional processes are derivatives of laser grooving, the physical mechanisms of the laser grooving process must first be understood before analyzing the more complex 3-D processes. The material removal rate for such processes is related to the depth of each groove. The groove depths define the boundary of the volume of material removed, and the scanning velocity defines the rate at which this volume is formed on the workpiece. The surface quality is related to the heat-affected zone and surface roughness at the groove surface during beam/material interaction. Dimensional accuracy is related particularly to the taper angle for each of the two grooves.

In the laser grooving process, a groove is produced by scanning a laser beam over the workpiece surface. Unlike through-cutting, the laser beam does not penetrate through the entire workpiece thickness. The physical mechanisms of laser grooving (Fig. 3.6) are similar to those of drilling and cutting.

Similar to laser grooving, laser scribing creates a blind groove on the surface of a workpiece. However, in laser scribing the ratio of groove depth to groove width is close to one, and the groove depths are typically very small.

Three-dimensional laser machining, laser grooving and laser scribing have the following advantages over conventional techniques:

- Three-dimensional laser machining can perform turning, threading, and milling operations on materials which are difficult to machine mechanically due to high hardness, brittleness, and abrasiveness.

- Lasers can be used to scribe or mark permanent identification patterns on metallic or ceramic parts. Laser-marked ID's can withstand greater amounts of wear than those marked with other methods.

- Since lasers can be focussed to a small spot, they are ideal for micromachining applications to repair defective integrated circuit components which would otherwise be scrapped.

On the other hand, three-dimensional laser machining, laser grooving and scribing have the following drawbacks:

- When applied to metals and ceramics, these processes result in molten material accumulation at the erosion front. However, unlike in laser cutting, a coaxial gas jet is not effective for ejecting molten material due to the presence of the groove bottom. The use of an off-axial gas jet, explained later in the chapter, can minimize the molten layer.

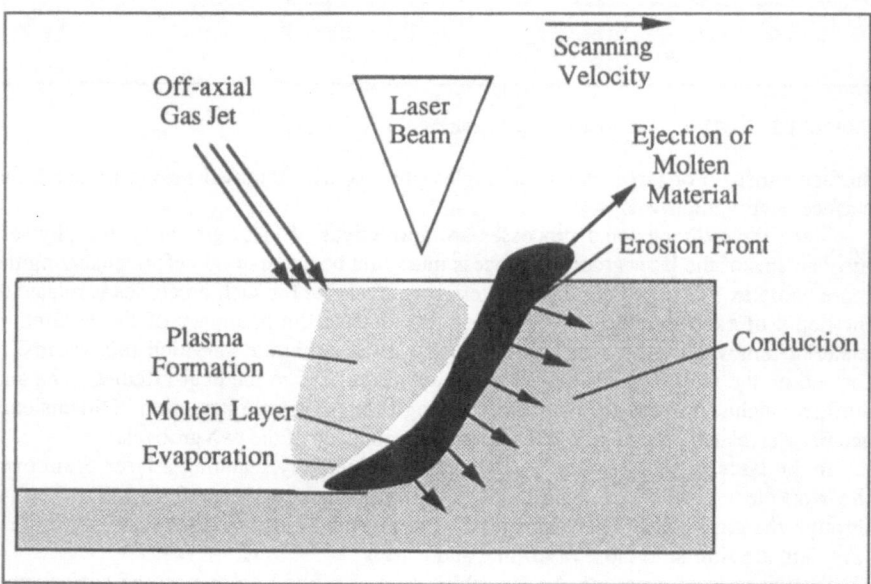

FIGURE 3.6   Laser Grooving

- The groove depth can fluctuate due to disturbances caused by laser beam changes, mechanical vibrations, material impurities, and gas jet fluctuations. In three-dimensional laser machining, unevenness in groove depth can decrease surface quality and the mechanical strength of the finished part. Consistency in groove depth can be maintained by using a closed-loop control scheme, which is also discussed later in this chapter.

3.2 Laser Machining Systems

The implementation of any laser machining process requires the integration of a number of optical, electrical and mechanical components into a machining system, which usually includes four major subsystems (Fig. 3.7):

- Laser beam generation
- Beam delivery
- Workpiece positioning
- Auxiliary Devices

Beam generation is accomplished within the laser device, which was discussed in detail in Chapter 2; in industrial practice, CO_2 and Nd:YAG lasers are most commonly used for beam generation in laser machining systems.

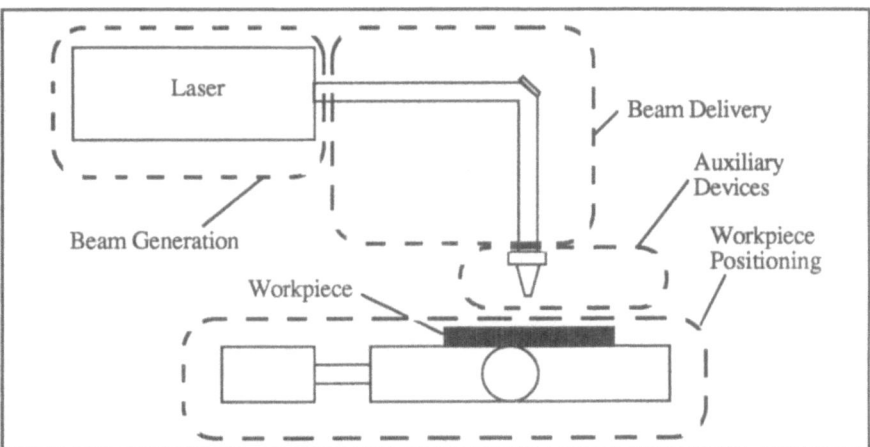

FIGURE 3.7   Components of a Laser Machining System

3.2.1   Beam Delivery

The beam delivery system consists primarily of optical components which focus the laser beam onto the workpiece surface. The major components of the beam delivery system are beam polarizers, mirrors and beam splitters, focussing lenses and fiber optic couplings. In a manufacturing environment, one or more lasers can be used in a time-share beam delivery system with several processing stations. In a time-share system, each laser and each processing station is connected to a central beam splitter or mirror via optical assemblies. The workpiece to be machined is placed at a processing station, where the desired laser type and beam power can be selected for a particular machining operation by switching the positions of the optics in the beam delivery system to connect the processing station with a laser station. This system is most suitable for situations in which setup time is significant compared to processing time. A list of vendors for beam delivery components is shown in Appendix A.

Circular Polarizers

As a photon travels in space, it has an oscillatory motion which causes oscillations in its electric field and produces an electromagnetic wave. This wave represents the direction and magnitude of the photon's electric field vector as a function of time (Fig. 3.8). In incoherent light, the photons oscillate with different frequencies, phases, and amplitude directions, which produces electromagnetic waves with random orientations and phases; due to the coherence and monochromaticity of laser light, the photons in the laser beam have identical amplitude, direction, and phase. Thus, the electric fields for all photons in the laser beam are aligned in the same direction; this property leads to a linear polarization of the laser beam.

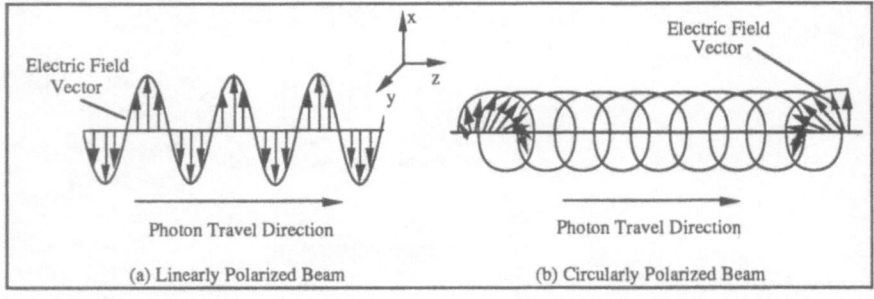

FIGURE 3.8   Linearly and Circularly Polarized Light

For cutting and grooving operations, the cutting speed, the width of cut, and the curvature of the kerf or groove depend upon the relationship between the amplitude direction and the beam scanning direction. A deep, narrow cut can be achieved when the electric field is oriented in the same direction as the beam scanning velocity, whereas a wide, shallow cut results from a 90° difference between the electric field vector and the scanning direction. Curvature of the groove can become visible at angles which differ 5° or more from the normal and parallel orientations, and the maximum curvature occurs at a 45° angle between the two orientations. The curvature effect is most apparent in a deep groove when the laser is at a high beam power and low scanning velocity.

Directional effects on cutting with a laser beam can be avoided by using a circular polarizer (Fig. 3.9). Usually, mirrors coated with a birefringent material (material with refraction index dependent on beam polarization direction) are used. The coating has an optical axis which is characteristic of the material used. Electromagnetic waves with polarization along the optical axis will pass through the material and waves with other orientations will be reflected. When a linearly polarized beam impinges on a mirror aligned 45° from the beam direction, the incoming laser beam is divided into two parts. A portion of the laser beam is reflected from the top surface of the birefringent material, while the remainder of the beam passes through the coating and is reflected by the top surface of the mirror material. After both portions are reflected from the mirror, they are rejoined; however, the electromagnetic waves which passed through the coating now possess an amplitude direction which is rotated 45° from the waves reflected at the coating surface. Also, the two electromagnetic waves are shifted out of phase by 1/8 of their wavelength. By using four mirrors aligned 45° from the incident beam, a circular polarizer can produce two linearly-polarized electromagnetic waves with a 180° phase difference. This causes a spin on the overall laser beam polarization direction, so that the preferred orientation of the electric field will vary in direction with time (Fig. 3.8). However, since the electromagnetic waves for all photons spin at the same rate, the waves still have identical direction and phase, so the properties of the laser beam are preserved. In machining operations, a circularly polarized beam shows no preference in cutting direction.

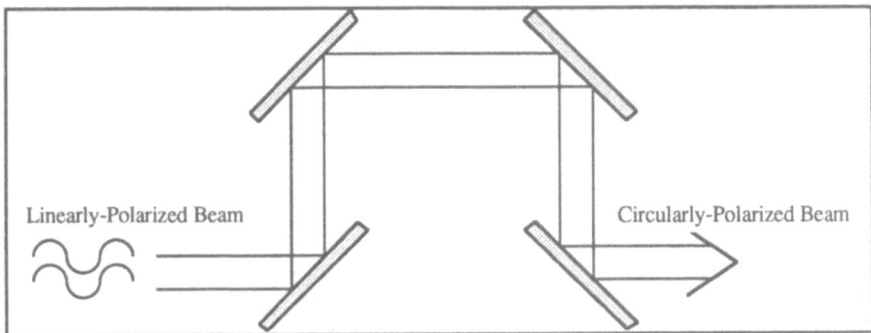

Linearly-Polarized Beam

Circularly-Polarized Beam

FIGURE 3.9   Schematic of Circular Polarizer

Mirrors

Metallic mirrors are generally used in beam delivery systems. Since metals have high reflectivities, the energy loss from the laser beam to the mirror is minimal. Metallic mirrors can also withstand high energy densities without thermal damage, because of the high thermal conductivity of most metals. The most common material for mirrors is copper, which can withstand energy densities above $100kW/cm^2$ without sustaining thermal damage. However, uncoated copper surfaces are soft, difficult to clean, and easy to oxidize and stain. A dielectric material is sometimes applied to the mirror to protect the surface and enhance its optical properties. Mirror surfaces are usually diamond-polished to achieve a flat surface. Water cooling is usually required to prevent excessive thermal distortions in mirrors.

Beam Splitters

A beam splitter is an optical device, usually flat, that is coated to reflect a portion of the incident beam energy and transmit the remainder. Beam splitters are used in many multi-workstation applications to distribute the output of a single laser among several stations. Standard reflection/transmission ratios used are 10/90, 30/70, and 50/50. There are three general methods for splitting a laser beam: reflection, interference, and polarization.

In the reflection method, a "roof splitter" composed of a copper wedge with a 90° edge angle can be used (Fig. 3.10). The edge faces directly into the laser beam, and part of the beam is reflected to either side of the wedge. By varying the wedge position relative to the laser beam, the ratio of the power levels for the two reflected beams can be varied. The copper wedge requires smooth polishing, a clean operating environment, and a high dimensional accuracy in the wedge geometry to avoid diffraction of the laser beam. The reflection method is highly dependent on the spatial mode and temporal stability of the incoming laser beam.

An alternative device to the roof splitter is an Inconel-coated ZnSe block (Fig. 3.10). The coating density can be varied on the block to provide either varying regions of reflectivity or a continuous gradient of increasing reflectivity. By targeting the incoming beam on different parts of the block, the reflection/transmission ratio can be changed.

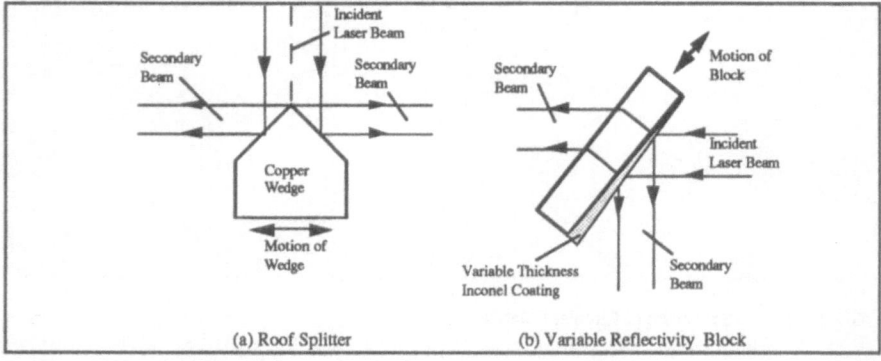

FIGURE 3.10   Beam Splitters Based on Reflection

The interference method of beam splitting uses a Fabry-Perot interferometer, which is composed of two parallel, closely spaced, partially-reflective ZnSe surfaces (Fig. 3.11). Depending on the beam wavelength and the separation distance between the plates, the waves of the photons from the reflected and transmitter beams will either destructively or constructively interfere. In destructive interference, each wave cancels the energy of the other, and 100% of the incoming beam is reflected; in constructive interference, the energy of the two waves are added, and all of the incoming beam is transmitted. This interference effect can be evaluated through a phase delay δ, defined as:

$$\delta = \frac{4\pi w \cos\theta}{\lambda}$$

(3-2)

where   w = plate separation distance
        θ = angle of incidence of incoming beam
        λ = wavelength of incoming beam

In destructive interference, the phase delay is 180°, while in constructive interference, the phase delay is 0°. The reflection/transmission ratio is related to the phase delay through the relationship:

$$\frac{I_T}{I_R} = \frac{(1 - r)^2}{(1-r)^2 + 4r \sin^2\left(\frac{\delta}{2}\right)}$$

(3-3)

where I_T is the transmitter beam intensity, I_R is the reflected beam intensity, and r is the surface reflectivity. By varying the plate separation distance w, the phase delay and reflection/transmission ratio can be changed. By using piezo-electric positioning devices, the reflection/transmission ratio can be changed in-process with a delay time as low as

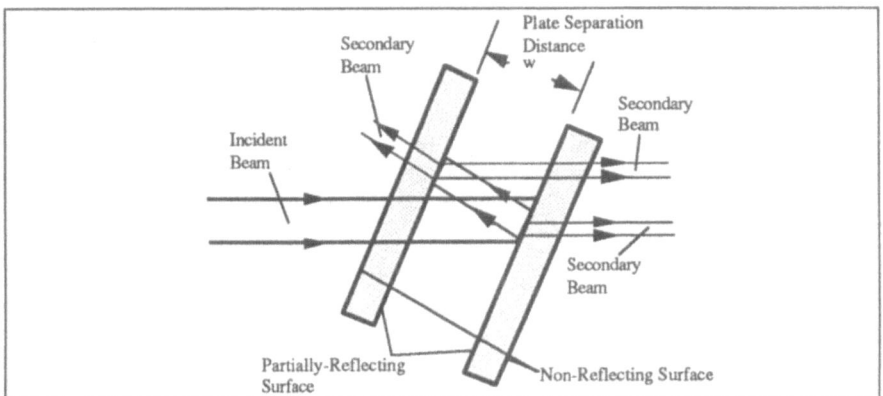

FIGURE 3.11   Fabry Perot Interferometer

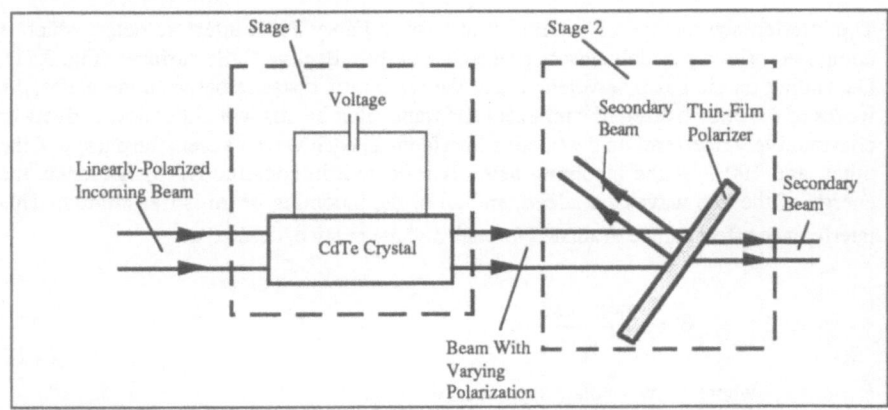

FIGURE 3.12   Polarization Beam Splitter

0.01s. However, the cost of a Fabry-Perot Interferometer is high compared to other methods, and the two plates require frequent and precise alignments.

The polarization method of beam splitting uses a two-stage process to vary the polarization of the incoming linearly-polarized beam and separate it into two beams (Fig. 3.12). The first stage is a Cadmium Telluride (CdTe) crystal which changes the orientation of the electric field vector depending on the voltage placed across the crystal. The second stage is composed of a thin-film polarizer which allows photons with electric field vectors parallel with the polarizer surface to pass through. As the voltage across the crystal is varied, the beam exiting the crystal will have an electric field vector orientation which varies between parallel (p-polarization) and perpendicular (s-polarization) to the polarizer surface. When this beam strikes the polarizer surface, all the photons with p-polarization are transmitted, while all photons with s-polarization are reflected.

Focussing Lens

A focussing lens is used to concentrate the laser beam into a small spot with high energy density. Sodium chloride (NaCl), potassium chloride, zinc selenide, gallium arsenide, and germanium are common lens materials for CO_2 and Nd:YAG lasers. The lens material must have high transmissivity at the laser light wavelength. The minimum focussed spot diameter, d, and the depth of focus, b, which defines a working range for laser machining, depend upon the focal length, wavelength and the unfocussed beam diameter (Fig. 3.13). For a TEM_{00} mode, the following relationships are applicable [10]:

$$d = \frac{4\lambda f}{\pi W}$$

(3-4)

$$b = \frac{2\lambda f^2}{\pi W^2}$$

(3-5)

where λ is the beam wavelength, W is the diameter of the unfocussed beam, and f is the focal length. By increasing the focal length, the working distance can be increased; however, the spot diameter also increases, thereby reducing the power density. For laser machining of thin workpieces, it is advisable to use a minimal focal length in order to decrease spot diameter and increase power density; this results in a maximum material removal rate with narrow hole, kerf or groove widths. If the laser must operate in many different applications, then a larger focal length lens can be used in conjunction with higher laser power in order to maintain a high power density.

There is a trade-off between spot diameter and working distance: a decrease in the spot diameter increases the energy density but decreases the effective working range. The spot diameter can be decreased either by decreasing the focal length or by increasing the unfocussed beam diameter. The unfocussed beam diameter can be increased by using a reflective or transmissive beam expander (Fig. 3.14).

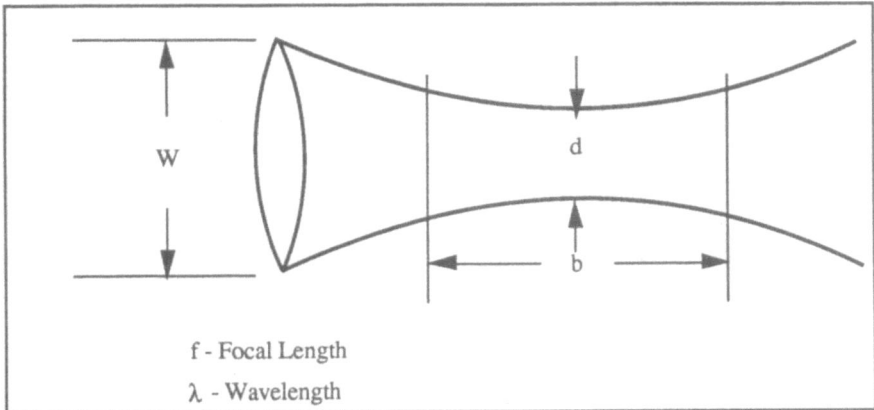

FIGURE 3.13   Focussing Lens Characteristics

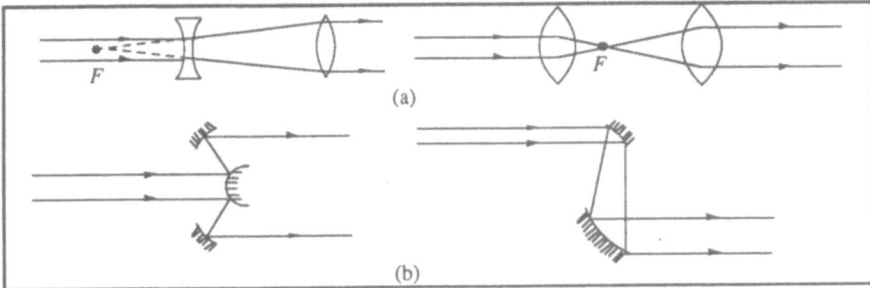

FIGURE 3.14   Beam Expanders (a) transmissive (b) reflective

Fiber Optic Coupling

One major problem with beam delivery systems based on mirrors is that geometric flexibility in laser nozzle positioning and beam aiming requires a complex collection of components. This complexity increases the difficulty of connecting lasers to devices having multiple degrees of freedom, such as robots; such devices are required to machine parts with complicated geometries. One alternative to using mirror assemblies for beam delivery is to use a fiber optic coupling between the laser output and the laser head. Single-mode (for TEM_{00} beam mode transmission) and multiple-mode (for higher-order mode transmission) optical fibers can be used for beam delivery. One problem with using fiber optics with industrial lasers is the limit on power that can be transmitted through them. The maximum laser power transmitted is limited by a dielectric breakdown at the entrance to the optical fiber. Beam power up to 100W (continuous wave) and 500W (pulsed) may be transmitted. A second problem is that a fraction of the beam energy is absorbed by the optical fiber material. Finally, beam transmission through optical fibers can cause mode distortion during beam delivery. For single-mode fibers, a Gaussian beam profile can be maintained throughout the fiber. For multiple-mode fibers, however, a higher order beam mode (such as TEM_{01}) will become distorted in bends of the fiber.

3.2.2   Workpiece Positioning

Two approaches can be used to control the motion between the laser beam and the workpiece:

- The workpiece can be moved using rotational and translational stages while the beam remains stationary (Fig. 3.15). This method is commonly used in current laser-cutting systems. Workpiece positioning allows a high degree of accuracy and high scanning velocities. For large workpieces, however, a substantial amount of floor space may be required. Also, workpiece positioning systems are limited by the complexity of part geometry. Workpiece motion is programmed using conventional NC languages (G or M codes). In some systems, the workpiece motions can be downloaded from a CAD/CAM system through an RS232 interface.

- The laser can be moved using translational and rotational stages while the workpiece is fixed (Fig. 3.16). Beam positioning allows the machining of intricate details and complex shapes, due to the more flexible kinematics of the nozzle compared to workpiece positioning systems. However, since the motion system must support the weight of the laser head, this scheme is limited in speed. An alternative is to use a flexible optical system or fiber optics to deliver the beam from a stationary laser to a position-controlled nozzle.

- A combination of workpiece and beam positioning can also be used in a coordinated fashion to machine intricate contours on parts (Fig. 3.17).

FIGURE 3.15   Laser Cutting System for Automobile Body Panels [28]

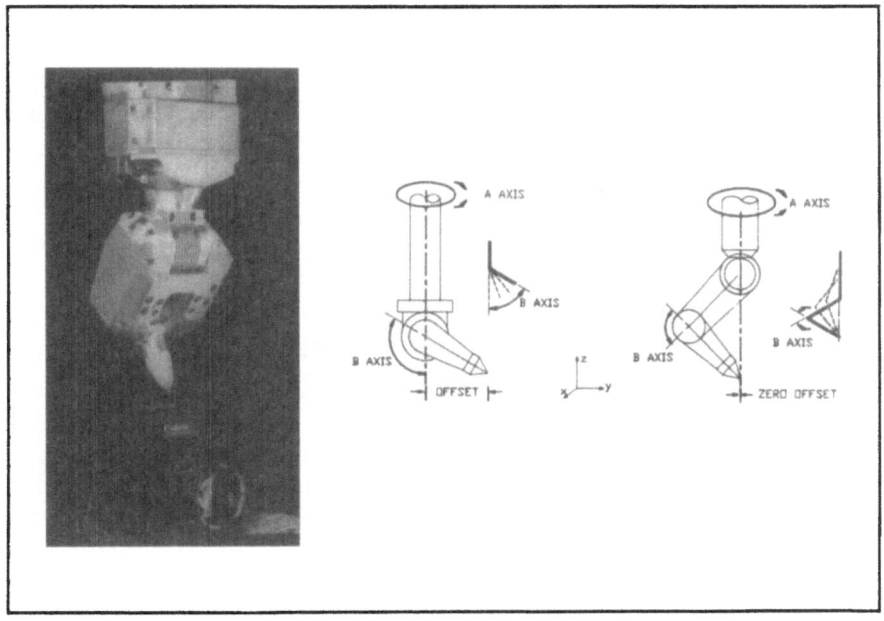

FIGURE 3.16   A Five-Axis Beam Positioning System [28]

Recently, a number of laser systems have been developed which incorporate multi-axis robotic manipulation of beam position. The robotic manipulation resulted from a need to perform complex laser welding, cutting and drilling on large workpieces. The first five-axis beam delivery systems were used for drilling cooling holes in turbine blades for aircraft engines [1]. Multi-axis workstations are also used for trimming panels. Five-axis control is achieved through three translational and two rotational degrees of freedom for the laser head. Six-axis beam delivery systems utilize articulate robots to direct the laser beam. Laser head motion for intricate shapes can be programmed by using a "teaching" mode, where a hand-held controller is used to direct the robot through each motion step. The criteria for selecting a multi-axis system are [24]:

- High accuracy
- High repeatability
- Smooth motions
- Movement and speed reference from the beam focal point
- Beam power control

The main advantage in using a multi-axis system is the reduction in floor space required for machining large workpieces (up to 40% workspace reduction over workpiece motion systems [1]). However, the accuracy of positioning and velocity of multi-axis systems is limited when machining workpieces with complicated geometries. Positioning errors result from mechanical vibrations, thermal expansion of the positioning links, and incorrect sensing of the laser head position compared with the position desired. Velocity inaccuracies are caused by the complicated kinematics of the positioning system. The number of calculations needed to maintain a constant velocity while maintaining the correct workpiece/laser head orientation may become excessive for a complicated part shape. Finally, the cooling requirements for optics in the beam positioning system may require the design of complex cooling systems.

FIGURE 3.17   A Three-Axis Laser Machining System With Both Beam and Workpiece Positioning
[28]

3.2.3   Auxiliary Components

Auxiliary components include equipment that is not used to direct the laser beam or position the workpiece. These components (gas jet nozzles, safety equipment, etc.) are used in support of the laser machining process.

Laser Head

The laser head is used to enclose the focussing laser beam and direct the coaxial gas jet towards the erosion front. The coaxial gas jet is used to prevent debris from impacting the focussing lens during laser machining. However, effective design of the nozzle and use of an assist gas also results in an improvement in surface quality.

Effective laser machining requires maximum expulsion of the molten material which is formed at the erosion front. The driving force for material expulsion is created by a pressure gradient at the erosion front resulting from the use of a co-axial or off-axial gas jet. In laser cutting, a coaxial jet is used to create a large pressure difference between the top and bottom of the kerf (Fig. 3.18). This pressure difference forms a downward driving force, which expels molten material through the bottom of the kerf. In laser grooving, an off-axial jet is used to create a pressure gradient along the erosion front (with high pressure at the bottom and low pressure at the top), as shown in Figure 3.18. This pressure gradient results in an upward driving force which expels molten material from the top of the groove. Therefore, in laser grooving, off-axial jets are more effective than co-axial jets in providing a driving force for molten material expulsion.

Most current laser systems use a coaxial nozzle with a parallel or convergent flow passage (Fig. 3.19). Several studies have been conducted in the use of coaxial gas jets in through-cutting and grooving processes. In [30], laser cutting with a variety of assist gas mixtures at high reservoir pressures was studied. A supersonic flow was achieved for air at reservoir pressures above 190kPa. However, a Mach shock disk may form at jet pressures above 350kPa; since the pressure drops discontinuously across this Mach shock disk, the laser machining effectiveness decreases. The occurrence of a Mach shock disk can be minimized by using a nozzle/workpiece distance above 3.5mm. In

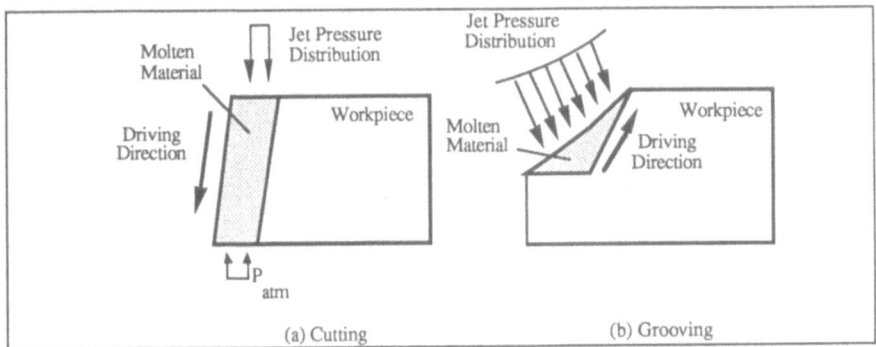

FIGURE 3.18   Erosion Front Pressure Distributions for Through-Cutting and Grooving

[14] and [38], supersonic jet flow from co-axial nozzles used in laser through-cutting was investigated. The convergent design was found to be most effective in through-cutting experiments for metal workpieces up to 3mm thick. The criteria used to judge for nozzle effectiveness are:

- Sharp corners at the kerf edges
- Smooth cut surfaces
- Parallel sides of the kerf
- Removal of molten material
- Narrow kerf width
- Minimal heat-affected zone

Recent experimental studies in nozzle design have used convergent-divergent and ring nozzles to improve laser cutting [30]. A convergent-divergent nozzle is used in conjunction with a supersonic jet to produce a more favorable shock structure for through-cutting (Fig. 3.20). Convergent-divergent nozzles minimize the sharp corners encountered on convergent jets, resulting in less divergence of the exiting gas. A ring nozzle allows the simultaneous flow of two gases through separate circular and annular sections (Fig. 3.20). With a convergent-divergent nozzle, clean kerfs can be made at higher cutting speeds that with a convergent nozzle. The pressure of each gas can be independently controlled to regulate the gas mixture. In cutting experiments on 5mm thick mild steel, the ring nozzle was more effective in removing dross and improving surface quality than a convergent nozzle for reservoir pressures between 3bar and 5bar and nozzle-workpiece distance between 1mm and 2mm [29].

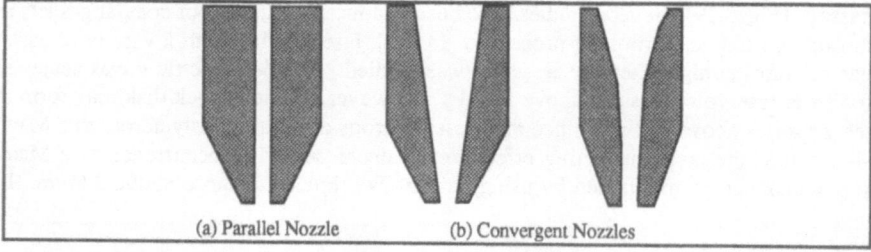

(a) Parallel Nozzle   (b) Convergent Nozzles

FIGURE 3.19   (a) Parallel and (b) Convergent Nozzle Designs

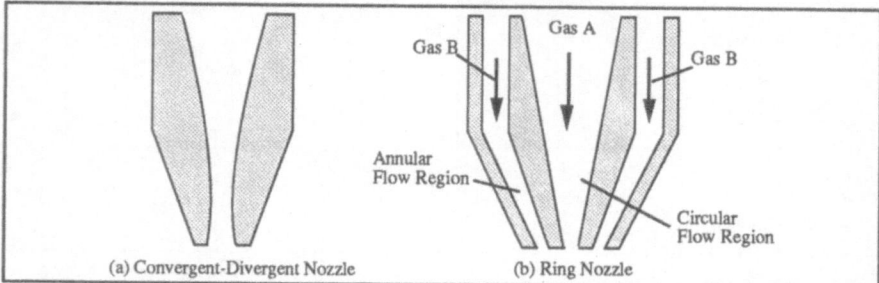

(a) Convergent-Divergent Nozzle   (b) Ring Nozzle

FIGURE 3.20   (a) Convergent-Divergent and (b) Ring Nozzle Designs

Nozzle exit shape can also have a substantial influence on gas jet effectiveness. In [14], 34 different non-circular nozzle exit geometries were investigated. Three lobed designs (Fig. 3.21) were found to be effective in delaying the formation of a Mach shock disk up to 6.5bar jet pressure. General guidelines for the design of coaxial nozzles for laser machining [39] are:

- Avoid the formation of a Mach shock disk in the cutting jet. The shock disk decreases laser machining effectiveness by decreasing the pressure difference between the top and bottom of the kerf or hole, thereby reducing the effectiveness of the jet in expulsion of molten material.

- Use workpiece pressure instead of nozzle pressure as a jet parameter, since a significant pressure drop can occur between the nozzle exit and the erosion front.

- Maintain a 3-4mm nozzle/workpiece distance to minimize the effect of a discontinuous pressure drop due to the formation of a Mach shock disk.

- Use a nozzle with a non-circular exit design (lobed exit) in order to operate at higher pressures without forming a Mach shock disk.

- Maintain tight quality control over machining of the nozzle exit. Any surface irregularities (such as burrs, notches, or eccentricity) will cause jet separation, which leads to a loss of pressure.

- Protect the nozzle from damage. Nicks and scratches on the nozzle exit can affect the jet flow.

In [8], optimal values for off-axial jet parameters resulting in high erosion front pressure gradients (Fig. 3.22) were determined. For grooving experiments performed on aluminum oxide, the use of an off-axial jet was found to improve groove depth by 20%.

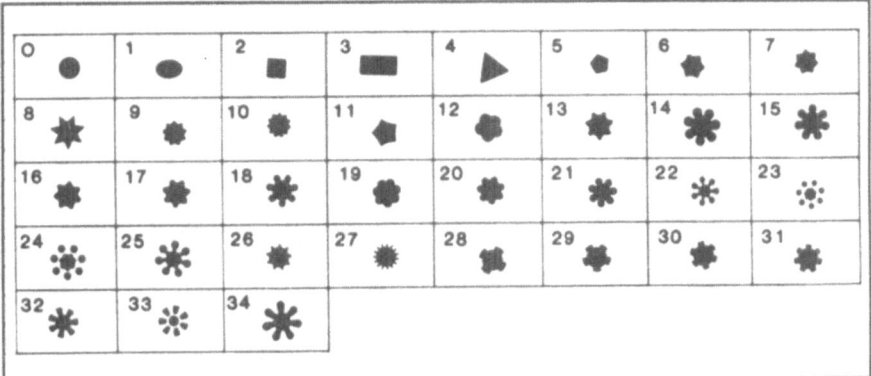

FIGURE 3.21   Non-Circular Nozzle Exit Designs [14]

The gas jet parameters found to significantly affect the pressure gradient included the reservoir pressure, nozzle-workpiece distance, jet targeting point, and jet attack angle. These parameters are related to the momentum transfer between the gas jet and the molten layer. A high reservoir pressure results in a large pressure difference between the bottom and top of the erosion front. The nozzle/workpiece distance affects the jet diameter at the kerf inlet. The jet targeting point determines which part of the jet and how large a portion of the jet will interact with the molten layer; it becomes critical when the erosion front is not entirely covered by the jet. In general, small nozzle/workpiece distances and large jet attack angles are the main jet characteristics which produce a large pressure gradient at the erosion front.

Laser Safety Equipment

Lasers pose a safety hazard due to the high energy densities present the beam and specular reflections, and the fumes that may be emitted by the workpiece during processing. All industrial lasers are classified as Class IV lasers [32] (hazardous to view under direct or diffusely-scattered conditions, fire hazard). The following precautions should be taken in a production environment:

- The beam delivery system should totally enclose the laser beam. Preferably, thick-walled tubes should be used to enclose the beam between the laser output to the laser head.

- The workpiece, laser head, and positioning stages should be enclosed to prevent the escape of stray beams. Ideally, the enclosure material should allow the work area to be viewed but should not transmit any of the laser energy.

Parameter	Value
Groove depth	<0.74 cm
Groove width	>0.08 cm and <0.13 cm
Groove angle	<30°
Jet attack angle	Perpendicular to the groove angle
Standoff distance	<0.12 cm
Jet directing point	>0.15 cm from the center in the opposite direction to the laser beam movement
Nozzle diameter	>0.10 cm and <0.20 cm

FIGURE 3.22   Gas Jet Parameter Values for Maximum Erosion Front Pressure Difference

- Doors for the laser processing areas should be interlocked to prevent entry or exit while the laser is on. The interlock will turn off the laser when a door is opened. Also, warning lights should be installed above doors to indicate when the laser is operating.

- Beam stops should be used around the workpiece to minimize beam reflection.

- An exhaust system should be constructed to remove harmful fumes from the laser processing area. Before attempting to machine a new material, experiments should be conducted to find out if the material produces harmful vapors when subjected to high temperatures.

- Warning signs should be prominently displayed in the work area. Class IV lasers require DANGER signs.

In some environments in which open beam processing is required, such as multi-axis workstations, it is necessary to define a nominal hazard zone (NHZ), where the possibility for potentially hazardous exposure exists [32]. For the direct unfocussed beam from a laser, the NHZ range is expressed as:

$$R = \frac{1}{\Phi} \sqrt{\frac{4P}{\pi ED} - W^2}$$

(3-6)

where ED is the unfocussed power density (W/cm^2), ϕ is the beam divergence angle, P is the laser power, and W is the unfocussed beam diameter (cm). For beam reflections from a laser-impinged diffuse surface such as the workpiece or beam stop, the NHZ range is given as:

$$R_d = \sqrt{\frac{\rho P \cos \theta}{\pi ED}}$$

(3-7)

3.3 Laser Machining Control

Since the capital and operating costs for laser machining systems are relatively high, high material removal rates, high dimensional accuracy, good surface quality and a high degree of repeatability must be achievable to make them economically viable. Currently, values for operating parameters for acceptable machining rate and quality are found through "trial and error" iterations, involving time-consuming calibration experiments to determine laser/material behavior. This procedure has several disadvantages. First, it must be repeated whenever a different workpiece material is machined, reducing the responsiveness of the laser as a machining tool. Second, using a trial-and-error method results in an acceptable machining condition, but not an optimal condition. Finally, there is no mechanism for reacting to process disturbances. Laser beam fluctuations, gas jet fluctuations, local changes in material composition, and velocity variations in the workpiece positioning system can lead to inconsistencies in depth of cut and surface quality.

In order to address these disadvantages, a closed loop control scheme can be used with the laser machining process. In this section, several possible methods for closed-loop control of laser machining are discussed. These include measuring the laser beam mode in-process to adjust the electrical current to the laser cavity, and control of beam power through direct observation of spark showers.

A method for regulating laser power was developed in [37] (Fig. 3.23). In this method, a laser beam analyzer provides a continuous measurement of beam spatial intensity distribution. A control signal is issued to the laser power supply to compensate for fluctuations in beam power and beam mode in order to maintain a constant energy density at the focal point.

A closed-loop control technique for optimizing surface quality for through-cutting of metals (Fig. 3.24) was implemented by [4]. Cutting geometry is inferred through optical observation and real-time image processing of spark shower patterns at the bottom of the kerf. A rule-based decision system is used to estimate cut quality from the image. The controller issues actuation commands to the laser power supply, gas flow regulator, and translation motors based on the cut quality estimate.

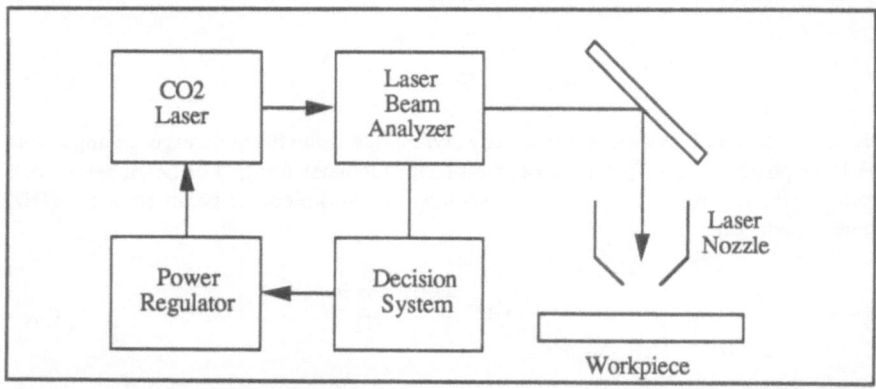

FIGURE 3.23   Laser Beam Regulation

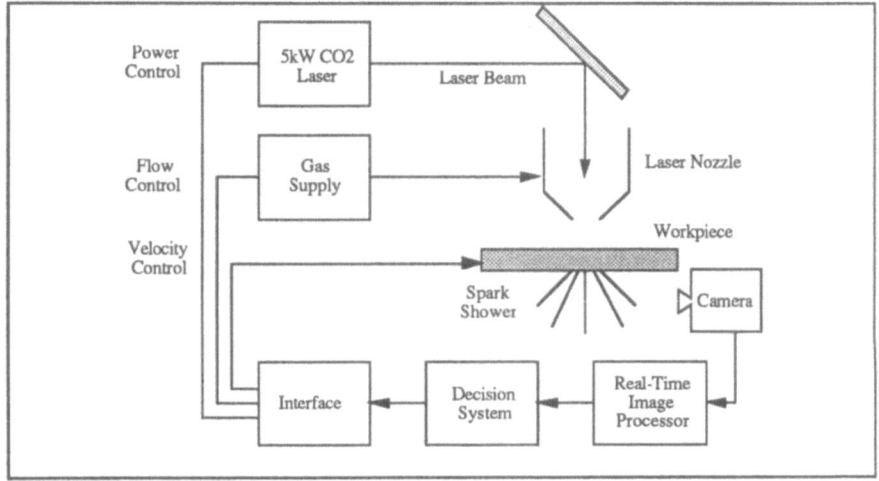

FIGURE 3.24   Kerf Regulation by Monitoring Spark Shower [4]

3.3.1 Sensing Devices

Accurate and reliable sensing of the condition of the erosion front by the control system is crucial. Possible techniques for real-time sensing of erosion front geometry during laser processing include laser beam analysis, direct optical sensing, ultrasonic pulse/echo, thermography, and acoustic emission. The advantages and disadvantages of each technique are discussed below.

First, *laser beam analysis* can be performed through continuous sensing of the spatial intensity distribution of the laser beam [2, 26, 35]. Local beam intensity and overall beam power can be measured by using a partially transmissive mirror, a rotating needle and detector, or a rotating hole and detector. The vibration of the reflector in the laser cavity can also be measured to determine power fluctuation. With these techniques, temporal power variations and beam fluctuations from the TEM_{00} mode to higher-order modes can be detected in-process. However, these techniques cannot detect other sources of disturbance to the beam/material interaction.

Direct optical observation was performed for laser through-cutting of steel in [11]. By using an endoscope, the kerf width and surface roughness at the kerf walls can be visually monitored. The disadvantage is that no information about depth of cut can be obtained from this observation. Groove depth can be inferred from optical observation of spark shower patterns (for metal cutting), but the relationship between spark shower patterns and depth of cut is very difficult to determine. Also, this technique will not be applicable for materials which do not produce sparks, such as ceramics and composites.

Ultrasonic pulse-echo techniques were used in welding to determine the shape of a weld pool [19] by transmitting a sound pulse through the workpiece and detecting the reflected pulse in order to determine the length of travel. This technique, however, is limited to simple workpiece geometries (such as rectangular slabs). Also, the large thermal gradients present in laser processing will significantly distort the ultrasonic pulses.

Thermography (infrared thermal imaging) has been used to monitor temperature distributions in welding [2] and laser cutting [3]. Thermography allows accurate sensing of high temperatures and large temperature gradients with fine resolution. Material impurities or fluctuations in groove shape can be detected by changes in isotherm shapes on a thermographic scan. However, the correlation between isotherms and groove shape is difficult to determine. Also, due to the changes in emissivity from melting of the material, the calibration of temperature values from measured radiation is difficult.

Acoustic emission measurements have been used in welding to detect crack initiation and propagation during cooling and in machining to determine tool wear [25]. An acoustic sensing technique has also been applied to laser machining for determining the cutting geometry with the aid of a gas jet [9]. During laser machining, a coaxial or off-axial gas jet is always used in tandem with the laser beam. The basic hypothesis of the acoustic sensing concept is that acoustic emission due to resonance is created when the gas jet impinges on the erosion front. The dynamic effects of jet flow in a constrained region such as the erosion front can be described by a three-dimensional wave equation. Vibrations of gas within a constrained volume will result in resonance at a characteristic frequency [33]. The resonant frequency varies depending on the cutting geometry. Therefore, by analyzing the frequency content of acoustic emissions sampled during laser machining, the cutting geometry may be determined.

The acoustic resonance from jet motion inside a cavity results from local compression and expansions in the constricted gas flow, which produces pressure waves. The acoustic wave propagation inside the kerf/groove is defined by the wave equation [3], which models the compression of a volume of ideal gas as sound waves pass through the volume:

$$\nabla \cdot \left(c^2 \, \nabla Q \right) = \frac{\partial^2 Q}{\partial t^2}$$

(3-8)

where c is the speed of sound through air and Q is the volume displacement of gas. The wave equation assumes ideal gas behavior and no heat interactions between the gas volume and the environment. Since the density of air at operating temperature and pressure ranges is low, ideal gas assumptions can be used. Also, since the heat transfer coefficient for convection to air is low [12], the convection effect from the erosion front to the gas jet is much less than the conduction heat into the workpiece, so the heat interaction between the erosion front and the gas jet can be ignored. During laser cutting or grooving, the gas jet expansion is constrained by the presence of the erosion front and the groove/kerf walls (Fig. 3.25). A *kerf* (open bottom) is produced in laser through-cutting and a *groove* (closed bottom) is produced in laser grooving. At the top surface, the jet cross-sectional area is assumed to be circular with a diameter equal to the groove/kerf width. Inside the groove or kerf, the jet is constrained by the side walls in the y direction and by the erosion front, but it is allowed to expand in the +x direction. If a linear expansion is assumed for the x direction with a slope of b, then the jet cross-sectional area can be expressed as a function of z.

$$A(z) = A_0 + A_1 z \quad \text{where}$$

$$A_0 = \frac{\pi}{4} w^2$$

$$A_1 = \frac{\pi}{4} wb$$

(3-9)

The jet expansion slope b has been experimentally determined as 0.097 through velocity measurements for the case of a free jet impinging on a slot [9]. Since the gas jet impingement on the erosion front of a groove or kerf is similar in configuration to a gas jet impinging on a slot, this jet expansion coefficient can be assumed for laser cutting and grooving. The groove/kerf width is assumed to be two times the focussed beam diameter based on analysis of experimental data. In order to proceed with the analysis, the following assumptions have been made (Fig. 3.26):

- Air motion is assumed to be unidirectional (z direction).

- The jet cross-sectional area is assumed to be proportional to the distance from the workpiece top surface ($A(z) = A_0 + A_1 z$).

- The air volume within the groove or kerf has been divided into differential elements. Air inside each differential element is assumed to have constant density and compressibility.

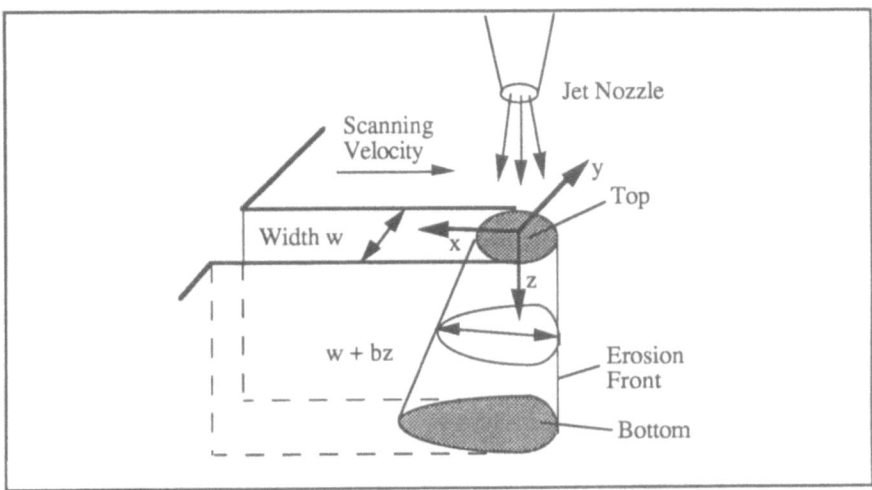

FIGURE 3.25   Cavity Cross-Sectional Geometry

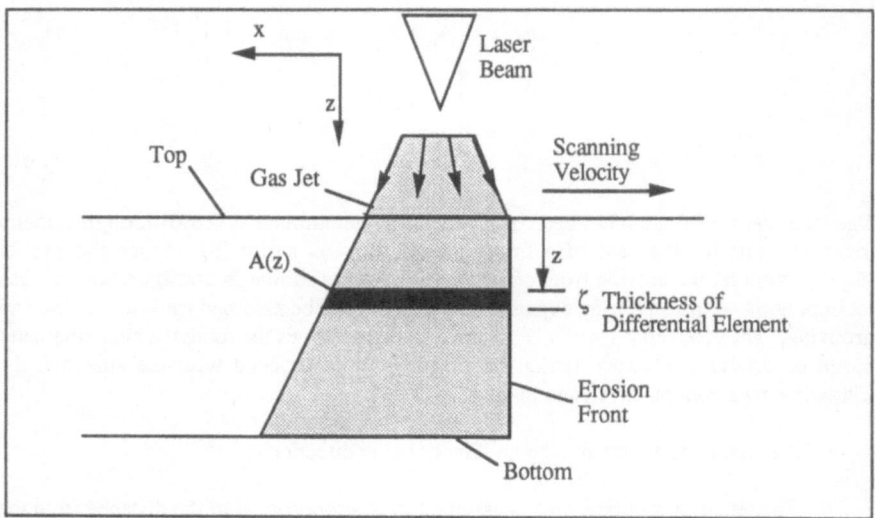

FIGURE 3.26   Differential Jet Element near the Cutting Front

With these assumptions, the wave equation for each differential element becomes:

$$c^2 \frac{\partial^2 \left(\zeta (A_0 + A_1 z) \right)}{\partial z^2} = \left(A_0 + A_1 z \right) \frac{\partial^2 \zeta}{\partial t^2}$$

(3-10)

where

$$Q = \left(A_0 + A_1 z \right) \zeta$$

(3-11)

Eq. (3-10) can be solved by using separation of variables and by considering the boundary condition: $\partial z / \partial z = 0$ at $z=0$ for all t. This boundary condition results from the assumption that there is no relative displacement of gas at the groove or kerf entrance and leads to the solution [9]:

$$\zeta(z,t) = C J_0 \left(\frac{\omega (A_0 + A_1 z)}{A_1 c} \right) \cos(\omega t)$$

(3-12)

where ω is the natural frequency of vibration for the differential element (in rad/sec). For laser grooving, the jet velocity at the bottom of the groove is zero in the z direction and results in the boundary condition: $\partial z / \partial t = 0$ at $z=D$ for all t. This boundary condition is satisfied if:

$$J_0 \left(\frac{\omega (A_0 + A_1 D)}{A_1 c} \right) = 0$$

(3-13)

Therefore, the resonant frequency $f_0 = \omega/2\pi$ can be found for grooving by solving Eq. (3-13):

$$f_0 = \frac{1.202 \, A_1 c}{\pi (A_0 + A_1 D)}$$

(3-14)

In laser cutting, a similar result can be found. Free jet expansion occurs at the bottom of the kerf (boundary condition: $\partial z/\partial z = 0$ at $z = D$ for all t). The relationship between the resonant frequency and kerf depth is:

$$f_0 = \frac{1.914 \, c A_1}{\pi (A_0 + A_1 D)}$$

(3-15)

In laser drilling, the gas jet is constrained by the hole wall once inside the hole, so the jet cross-sectional area remains constant ($A_1 = 0$). By applying constant jet area to the three-dimensional wave equation (Eq. 3-8), a one-dimensional differential equation is established with the form:

$$c^2 \frac{\partial^2 \zeta}{\partial z^2} = \frac{\partial^2 \zeta}{\partial t^2}$$

(3-16)

Solution of Eq. (3-16) using the boundary conditions:

$$\frac{\partial \zeta}{\partial z} = 0 \quad \text{at} \quad z = 0 \quad \text{for all t}$$

$$\frac{\partial \zeta}{\partial t} = 0 \quad \text{at} \quad z = D \quad \text{for all t}$$

(3-17)

yields a relationship between resonant frequency and hole depth.

$$f_0 = \frac{c}{4D}$$

(3-18)

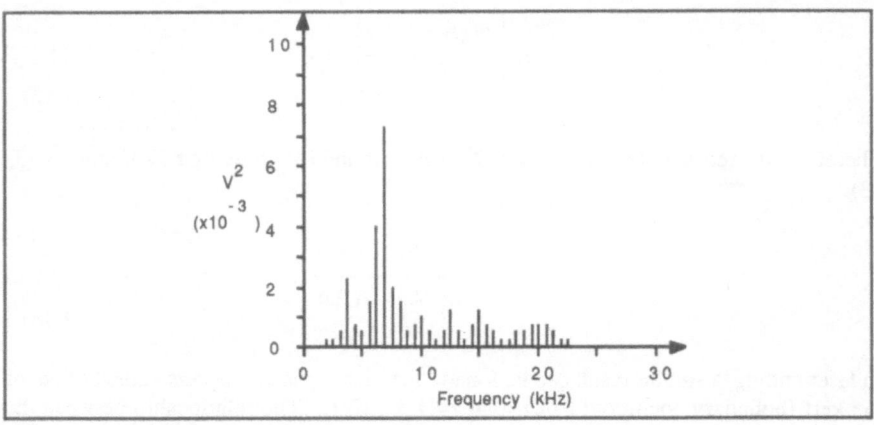

FIGURE 3.27   Frequency Spectrum for Laser Grooving (P=400W, V=1.5cm/s)

The frequency spectrum measured during laser grooving showed a resonant peak intensity more than two times larger in magnitude than intensities for other frequencies in the power spectrum (Fig. 3.27). The range of detectable frequencies was determined by the minimum and maximum frequency values, which exhibit acoustic intensity levels above the detected background levels ($6 \times 10^{-4} V^2$ for frequencies under 50kHz).

A comparison between the analytical estimate and experimental data (Fig. 3.28) shows that Eq. (3-14) overestimates groove depth compared with experimental results. Since the resonant frequency is less sensitive to groove depth variations at large depth values, the error for analytical estimates increases from 22% at a groove depth of 5mm to 42% at a 13mm groove depth. The error on the groove depth estimates can be attributed to the transverse component of jet velocity (velocity in the x direction in Figure 3.25) due to the presence of an inclined erosion front and the groove bottom, which deflects jet flow to the transverse direction (Fig. 3.29). The analysis assumed a vertical erosion front and a rectangular cross-section with vertical groove walls and a flat groove bottom (Fig. 3.30); actual groove cross-section shows a tapered groove with a sharp groove bottom, which also contributes to creating transverse flow and jet mixing inside the groove. The presence of transverse flows inside the groove results in a greater expansion of cross-sectional area of the gas jet inside the groove compared with analytical estimates. The transverse flow and tapered groove shape cannot be considered analytically with Eq. (3-8), so the assumptions are necessary in order to derive an analytical relationship useful for implementation in a control system. As the groove depth increases, the amount of jet mixing and transverse flow inside the groove also increases. Therefore, the analysis assumptions may have a significant effect on the accuracy of analytical estimates for groove depths larger than 10mm. The measured frequency spectrum during laser through-cutting showed, similar to the grooving results, a sharp peak at the resonant frequency.

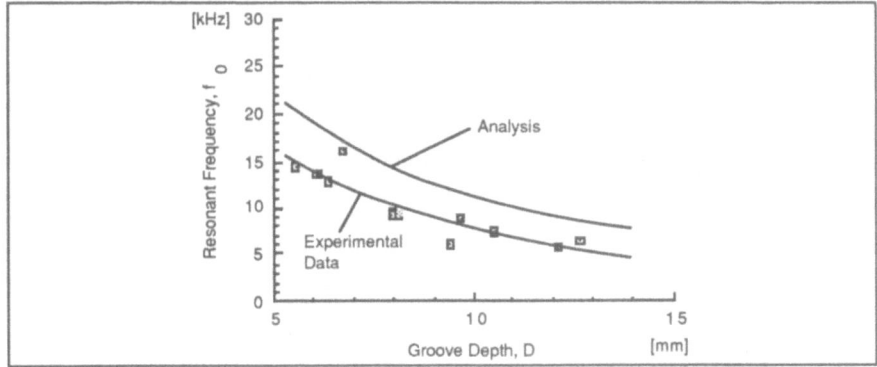

FIGURE 3.28   Resonant Frequency vs Groove Depth for Laser Grooving

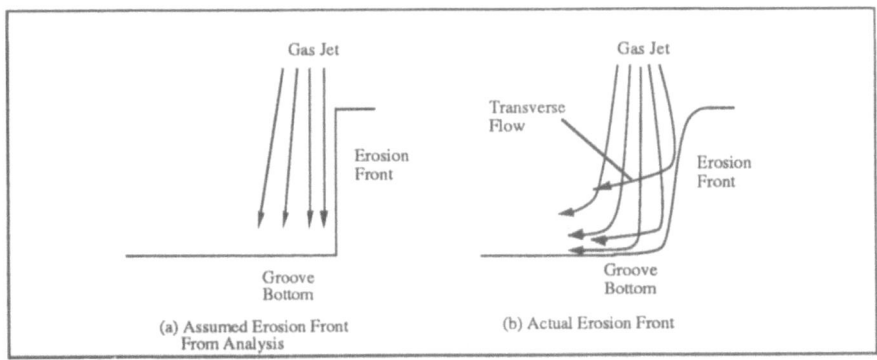

FIGURE 3.29   Difference Between Assumed and Actual Erosion Front

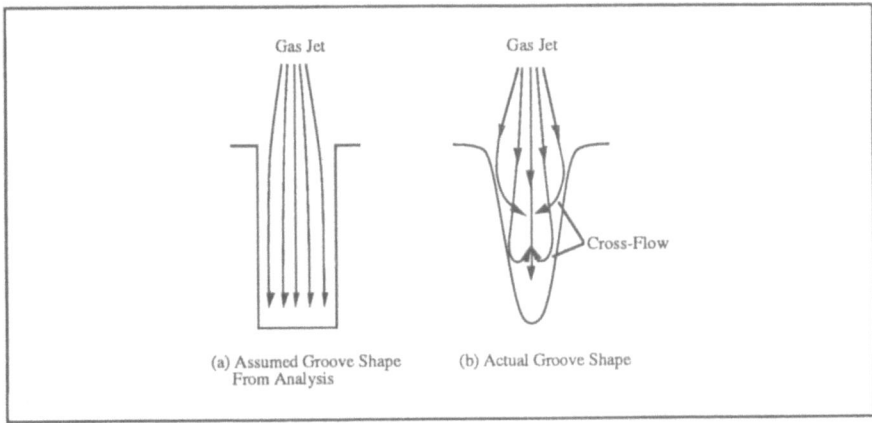

FIGURE 3.30   Difference Between Assumed and Actual Groove Shape

For laser cutting, the acoustic analysis overestimated the cutting depth from 20% at 23mm workpiece thickness to 44% at 2mm workpiece thickness (Fig. 3.31). Unlike in laser grooving, the deviation between analytical estimates and test results decreased as cutting depth increased. Possible sources of discrepancy between model estimates and test results were the effects of cross-flows in the kerf due to an inclined erosion front (resulting in a larger expansion of the gas jet in the kerf than estimated in the analysis) and the effects of jet flow around sharp corners at the kerf exit, which produced jet mixing due to flow separation (Fig. 3.32). As the workpiece thickness is increased, the corner flow effects decrease compared with the main unseparated flow, so the flow assumption made in the analysis becomes more accurate and the cutting depth error decreases.

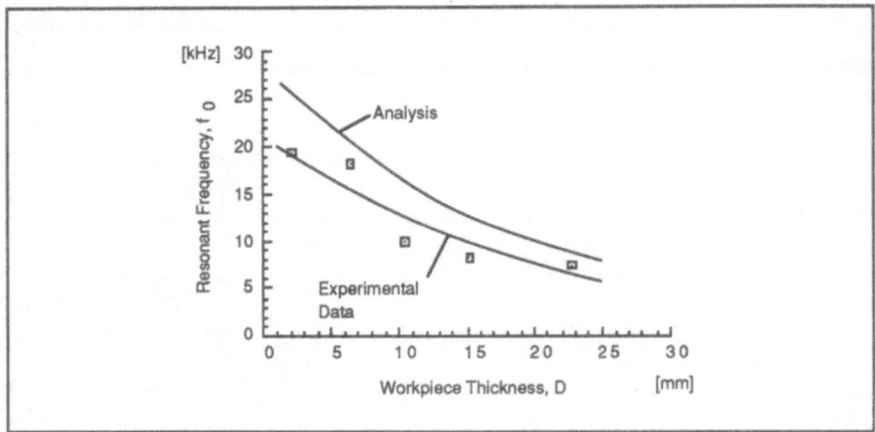

FIGURE 3.31   Resonant Frequency vs. Workpiece Thickness for Laser Cutting

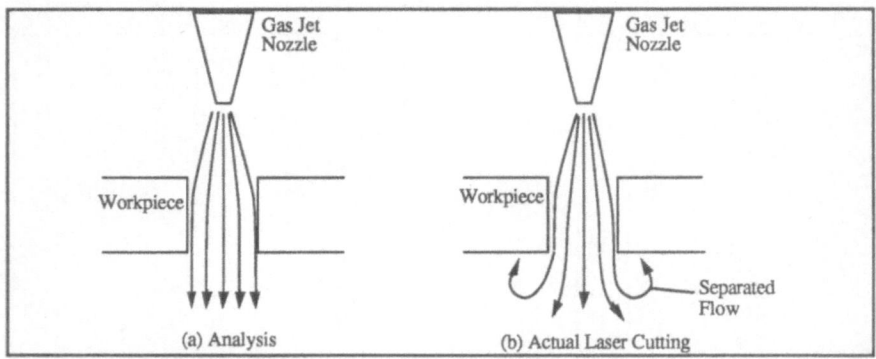

FIGURE 3.32   Effect of Jet Mixing at the Kerf Exit

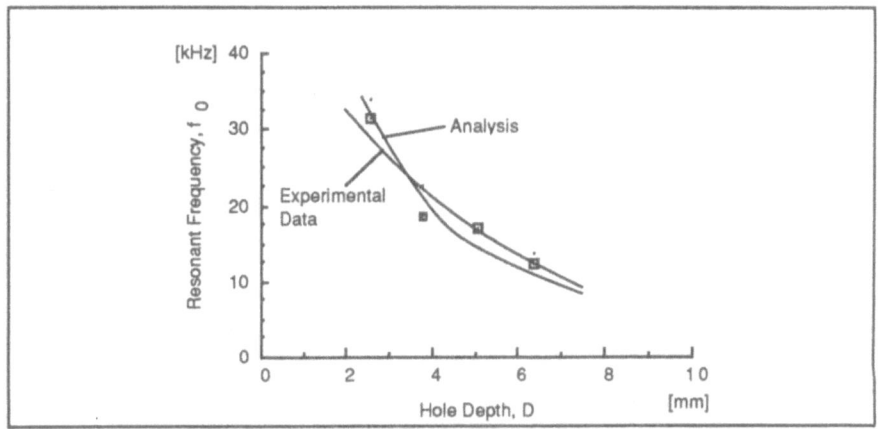

FIGURE 3.33   Resonant Frequency vs. Hole Depth for Laser Drilling

In the case of blind drilling, where the hole does not penetrate through the entire thickness of the workpiece, the analytical estimates of hole depth showed a small deviation (less than 15%) from experimental results over the entire test range (Fig. 3.33). The gas flow in the hole was constrained by the hole walls on all sides, which was accounted for by the analysis. Since no assumptions relating to gas jet expansion were made, the analytical estimates more accurately reflected actual test results.

Evaluation of the results for acoustic measurement in laser grooving, cutting, and drilling shows that the acoustic model can predict most accurately in the case of laser drilling. The analysis overestimates depth of cut for both laser grooving and cutting; however, the error between analysis and experimental results increases with increasing groove depth in laser grooving (due to jet mixing in the groove), while the error decreases with increasing depth of cut for laser cutting (due to jet separation at the open bottom). In order to improve the accuracy of the sensor model, the analytical results can be multiplied by a coefficient B and expressed as:

$$f_0 = \frac{A_1 B c}{\pi (A_0 + A_1 D)}$$

(3-19)

where B=0.88 for laser grooving and B=1.44 for laser cutting.

3.3.2 Future Directions for Laser Machining Control

Assuming that a laser machining process must fulfill certain requirements in material removal rate, surface quality, and dimensional accuracy, one or more state variables (depth of cut, heat-affected zone, and taper) can be defined for input into the control system (Fig. 3.34). Sensors are used to monitor process conditions through measured variables such as acoustic emission, temperature, or optical measurements. These

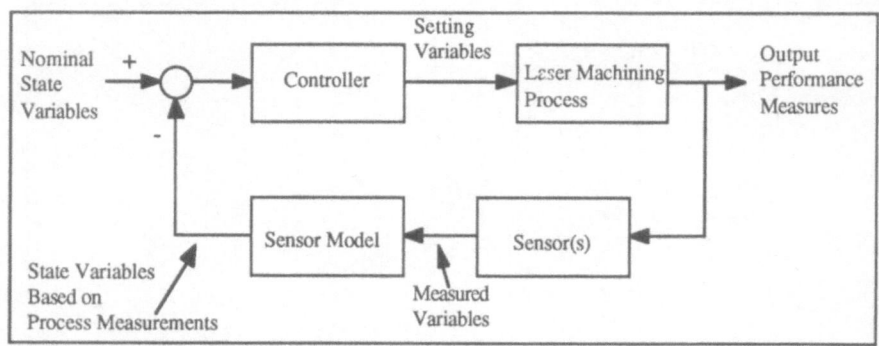

FIGURE 3.34   General Components for a Closed-Loop Control Scheme

measured variables are related to actual state variables through a set of sensor models. A comparison of actual and nominal state variables results in an error, which becomes the input into the controller. The controller changes the setting variables, such as the laser cavity current, gas jet parameters, and scanning velocity, in order to reduce the error. The controller parameters can be determined through simulation with the aid of a control model of the laser machining process.

Of the three laser machining processes, laser grooving presents the most general case in terms of analyzing control issues. Since the beam does not penetrate the entire material thickness, disturbances to the system may induce variations in depth of cut. Also, since laser grooving occurs in steady-state, a process sensor can reliably give continuous feedback of the erosion front condition. Therefore, the process control scheme in this section is simulated for the case of laser grooving. However, the process control presented here can also be applied to laser cutting and drilling.

The acoustic sensing technique can be integrated into a control scheme for laser grooving [9], as shown in Figure 3.35. In this control scheme, acoustic emission from the process is measured by the acoustic sensor. The sensor model (based on Eq. 3-16) converts the measured variable (resonant frequency) into a state variable (groove depth). A comparison between the estimated and nominal groove depth yields an error, which becomes the input to a proportional-integral-derivative (PID) controller. The controller relates the error to a change in the setting variable (scanning velocity). The improvements in groove depth fluctuations due to disturbances to the system and the dynamic response of the control system in selecting a setting variable value corresponding to a desired groove depth can be analyzed by performing simulations of the proposed control scheme.

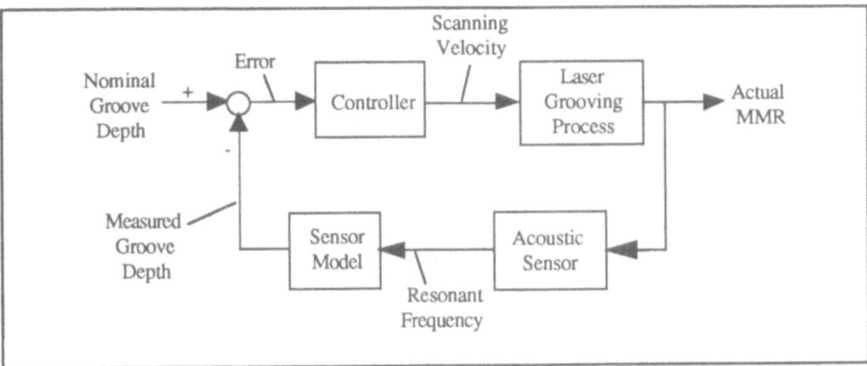

FIGURE 3.35   Control Scheme for Laser Grooving

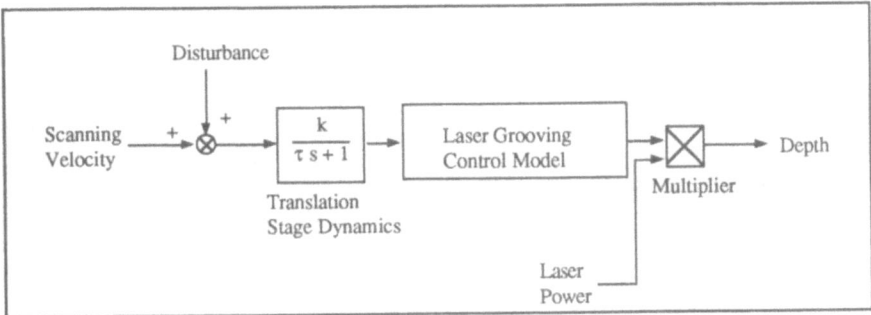

FIGURE 3.36   Block Diagram for Open-Loop Laser Grooving

For simulation of the open-loop system (Fig. 3.36), the laser grooving process was dynamically modeled as a first order system with the scanning velocity as input. Laser power affected system response through changes in gain for the process model. The workpiece translation stage was assumed to have a first order behavior. The process model used was an analytical solution derived from heat transfer considerations in [12, 13], which provided a relationship between the state variable (groove depth) and the setting variable (scanning velocity) through heat transfer analysis both at the groove surface and in the interior of the workpiece.

$$D = \frac{2aP}{\pi^{1/2}\rho V d\left(c_p(T_{vap} - T_\infty) + L\right)}$$

(3-20)

For the closed-loop system (Fig. 3.37), the acoustic emission was measured by the acoustic sensor and related to measured groove depth through the acoustic model. The error between the nominal and measured groove depths was an input signal to a controller with proportional, integral and derivative components (PID). The PID controller related the error signal to changes in the setting variable (scanning velocity). In order to model the control scheme, the following assumptions were made:

- The dynamics of the translation stage can be modelled as a first-order equation and higher-ordered dynamics can be neglected.

- The dynamics of the acoustic sensor are neglected.

- The PID controller has an initial condition of zero.

- The laser grooving process shows a linear relationship between the setting variable and the state variable.

- All fluctuations affecting groove depth (beam power fluctuations, gas jet fluctuations, vibrations in the translation stage, changes in material composition, etc.) can be lumped into one disturbance modelled as random noise with a magnitude of one standard deviation of experimental data.

From numerical simulations performed on the open-loop system, the system response to a step increase in velocity (Fig. 3.38) showed the groove depth reaching a steady-state condition at t = 0.1s without closed-loop control. With closed-loop control, the time required for depth to reach steady-state was reduced by a factor of ten to 0.01s (Fig. 3.38). The steady-state open-loop response (Fig. 3.39) showed groove depth variations of 4% of nominal value due to disturbances to the system. The closed-loop system was less sensitive to the effects of external disturbances, with groove depth variations of less than 0.5% of nominal value.

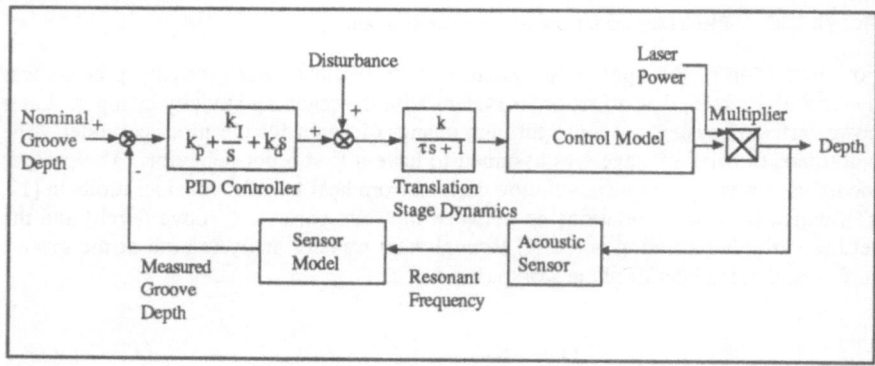

FIGURE 3.37   Block Diagram for Closed-Loop Laser Grooving

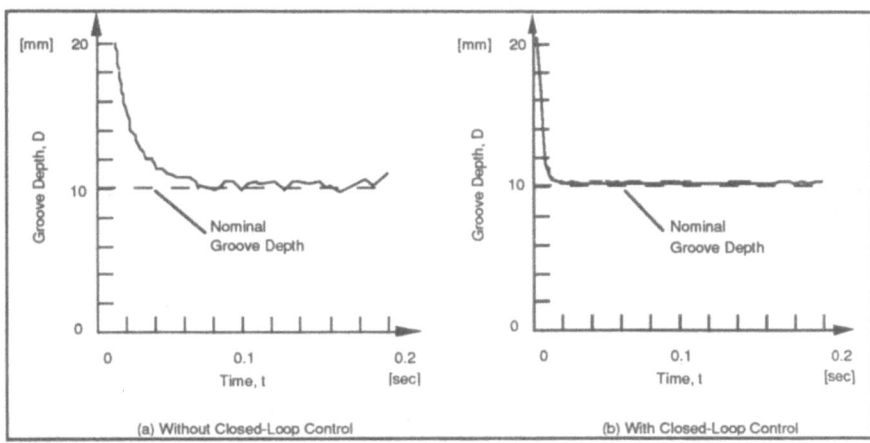

FIGURE 3.38   Groove Depth Response for Step Input (P=300W, Material: Acrylic)

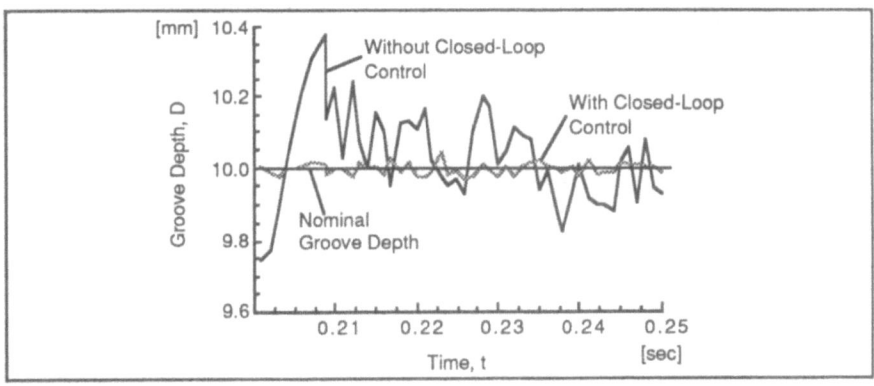

FIGURE 3.39   Steady-State Groove Depth Fluctuations With and Without Closed-Loop Control (Power=300W, Material: Acrylic)

Simulations were also performed with step increases in laser power to determine the response of the closed-loop system to laser power fluctuations (Fig. 3.40). For each increase in beam power, the control system compensated by increasing the scanning velocity to maintain a constant depth and maximize material removal rate. A steady-state error of 0.5% of nominal groove depth was maintained at each beam power level.

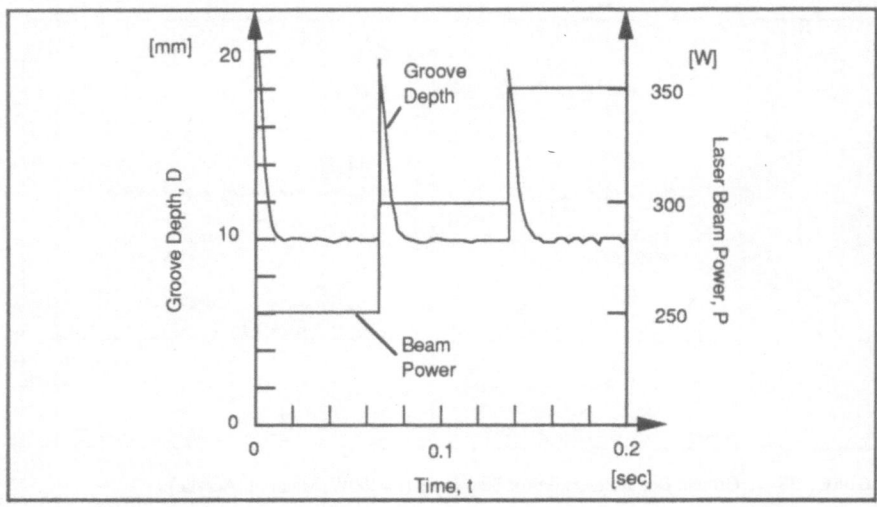

FIGURE 3.40   Depth Response to Beam Power Changes for Closed-Loop System

3.4 Economics of Laser Systems

In an economic context, lasers provide the following advantages over other processes:

- Laser machining eliminates the replacement costs associated with tool wear and breakage.

- Machine downtime for tool replacement and recalibration are eliminated.

- Lasers can be integrated in a computer numerical control (CNC) system to maximize machining rates.

- Due to the high precision of laser machining, closer nesting of parts can be accomplished, so less material is wasted.

- Unlike other nontraditional machining processes, laser machining does not require a vacuum or any other special environment to operate.

Laser systems have the following cost components:

- High capital expenditure is required. High-power lasers can cost up to several hundred thousand dollars. Most CO_2 lasers cost between $70 and $100 per watt [17]. Nd:YAG lasers are 10% to 20% higher on average. Other associated costs, such as the cooling system, power supply, beam delivery system, safety equipment, and workpiece positioning system or multi-axis robot may exceed the cost of the laser.

- Associated operating costs for a laser system include operating gas usage (He, N_2, and CO_2 gases for CO_2 lasers), electrical power usage, assist gas usage, and cooling water.

- Skilled operators are required to operate and maintain the system. Since the laser system is continuously operating in most industrial applications, full-time operators are required.

- The cost of purchasing and installing safety devices such as fume hoods, warning lights, goggles, beam enclosures, and door interlocks must also be accounted for. Installation of some safety devices may require major building renovation of the laser facility.

In general most metals and nonmetals have thermal and optical properties favorable for laser machining. However, the economic viability of applying laser machining depends critically on properly matching the material with the laser type and size. The absorptivity of a material is dependent on the wavelength of the incoming laser beam. The thermal properties of a material (melting/vaporization temperature, thermal conductivity, specific heat, and latent heat) determine the power requirement for an incident beam in order to achieve a desired depth of cut or material removal rate. Since ceramics, plastics, and composites have high absorptivity values for 10.6 micron radiation, they can be machined effectively with a CO_2 laser. Ceramics and composites generally have higher melting and/or vaporization temperatures, specific heats and latent heats than plastics; therefore, higher laser power is required. Some metals such as aluminum, copper, and brass exhibit low absorptivity values for 10.6 micron radiation, so most of the CO_2 laser energy will be reflected away from the workpiece surface. These materials are more suitable for machining with a Nd:YAG laser operating at a 1.06 micron wavelength. Materials with high reflectivity in both laser wavelengths, such as gold and silver, are difficult to machine with a laser.

The economic feasibility of implementing a laser system should be evaluated in the context of the machining application. For many difficult-to-machine materials, mechanical machining techniques cannot consistently produce high-quality parts. In some applications, laser machining can also achieve higher material removal rates than other mechanical or nontraditional machining methods. Finally, when used in conjunction with a multi-axis robot or a workpiece positioning system, laser machining is a highly flexible and programmable tool.

References

1.   Belforte, D.A., "Robotic Manipulation for Laser Processing," *Proceedings of the SPIE – High Power Lasers and Their Industrial Applications,* Vol. 650 (1986), 262-270.

2.   Beyer, E., G. Herziger, R. Kramer, and P. Loosen, "A Diagnostic System for Measurement of the Focused Beam Diameter of High Power CO_2 Laser," *Proceedings, SPIE Conference on High Power Lasers and Their Industrial Applications,* Vol. 605 (1986), 170-177.

3.   Boillot, J.P., et al., "Adaptive Welding by Fiber Optic Thermographic Sensing: An Analysis of Thermal and Intrumental Considerations," *Welding Research Supplement* (July 1985), 209-217.

4.   Burg, B., E. Lamotte, L. Foulloy, and B. Zavidovique, "A Smart Laser Cutter," *Proceedings of the SPIE – High Power Lasers and Their Industrial Applications,* Vol. 650 (1986), 271-278.

5.   Chin, B.A., N.H. Madsen, and J.S. Goodling, "Infrared Thermography for Sensing the Arc Welding Process," *Welding Journal* (Sept. 1983), 227-234.

6.   Chryssolouris, G., J. Bredt, and S. Kordas, "A New Machine Tool Concept Based on Lasers," *Proceedings, 14th NAMRC* (1986), 244-251.

7.   Chryssolouris, G., J. Bredt, S. Kordas, and E. Wilson, "Theoretical Aspects of a Laser Machine Tool," *ASME PED-20, Manufacturing Simulation and Processes* (1987), 177-190.

8.   Chryssolouris, G. and W.C. Choi, "Gas Jet Effects on Laser Cutting," *SPIE CO_2 Laser Technology and Applications Conference* (July 1988)

9.   Chryssolouris, G. and P. Sheng, "Process Control of Laser Grooving Using Acoustic Sensing," *Proceedings, ASME Winter Annual Meeting* (1990).

10.   Coherent, Inc., *Lasers – Operation, Equipment, Application, and Design,* Coherent Inc., McGraw-Hill, New York, 1980, 137-196.

11.   Decker, I., M. Hansmann and J. Ruge, "Facilities of Quality Control in Laser Cutting," *Proceedings, SPIE Conference on High Power Lasers and Their Industrial Applications* Vol. 605 (1986), 279-284.

12.   Duley, W.W., *Laser Processing and Analysis of Materials*, Plenum Press, New York, 1983.

13.   Eberhardt, G., "Survey of High Power CO_2 Industrial Laser Applications and Latest Laser Developments," *Proceedings of the First International Conference on Lasers in Manufacturing* (Nov. 1983), 13-19.

14.   Fieret, J., and B.A. Ward, "Circular and Non-Circular Nozzle Exits from Supersonic Gas Jet Assist in CO_2 Laser Cutting," *Proceedings, Third International Conference on Lasers in Manufacturing* (1986), 45-54.

15.   Floux, F., "Application of High Power Lasers to the Creation of Active Plasmas," *Optics and Laser Technology* (Apr. 1974), 69-77.

16.   Gukelberger, A., "Industrial Applications of High Power CO_2 Lasers - System Descriptions," *Proceedings of the SPIE - High Power Lasers and Their Industrial Applications,* Vol. 650 (1986), 254-261.

17.   Gregson, V.G., "Economics of Industrial Laser Processing," *Industrial Laser Annual Handbook,* Penwell Pub., Tulsa, OK, 1987, 31-37.

18.   Hadisty, F. B., "Development of a Multi-Workstation Laser Processing Facility," *Proceedings of the First International Conference on Lasers in Manufacturing* (Nov. 1983), 63-70.

19.   Hardt, D.E., and J.M. Katz, "Ultrasonic Measurement of Weld Penetration," *Welding Research Supplement* (Ept. 1984), 273-281.

20.   Heglin, L.M., "Introduction to Laser Drilling," *Industrial Laser Annual Handbook,* Penwell Pub., Tulsa, Oklahoma, 1986, 116-120.

21.   Herziger, G., "Physics of Laser Materials Processing," *Proceedings of the SPIE – High Power Lasers and Their Industrial Applications,* Vol. 650 (1986), 188-194.

22.   Hugenschmidt, M., "Interaction of Repetitively Pulsed High Energy Laser Radiation With Matter," *Proceedings of the SPIE – High Power Lasers and Their Industrial Applications,* Vol. 650 (1986), 195-201.

23.   Janjua, M.S., K. Rathmill, and D.M. Allen, "A Technical and Financial Appraisal of NC Laser Machines in a Profile Cutting Application," *Proceedings of the First International Conference on Lasers in Manufacturing* (Nov. 1983), 41-52.

24.   Johnson, T.A., "Robot Controlled Laser Processing," *Proceedings of the First International Conference on Lasers in Manufacturing* (Nov. 1983), 71-78.

25. Kannatey-Asibu, E. and D. Dornfeld, "Quantitative Relationships for Acoustic Emission from Orthogonal Metal Cutting," *ASME Journal of Engineering for Industry* (Aug. 1981), 330-340.

26. König, W., F.U. Meis and C. Schmitz-Justen, "Process Monitoring of High Power CO_2 Lasers in Manufacturing," *Proceedings of the Second International Conference on Lasers in Manufacturing. (LIM2)*, (1986), 129-140.

27. Laos, O.V., "Evaluating a CO_2 Industrial Laser System," *Proceedings of the First International Conference on Lasers in Manufacturing* (Nov. 1983), 21-30.

28. Laser Lab, Inc., product literature (1990).

29. Luxon, J.T. and D.E. Parker, *Industrial Lasers and Their Applications*, Prentice-Hall, Engelwood Cliffs, NJ, 1985, 200-242.

30. Nielsen, S.E., "Laser Cutting With High Pressure Cutting Gases and Mixed Cutting Gases," PhD Thesis, Technical University of Denmark, 1985.

31. Ready, J.F., "Materials Processing–An Overview," *Proceedings of the IEEE*, Vol. 70, No. 6 (June 1982), 533-544.

32. Rockwell, R.J., "Fundamentals of Industrial Laser Safety," *Industrial Laser Annual Handbook*, Penwell Pub., Tulsa, OK, 1987, 131-148.

33. Schuocker, D., "Laser Cutting," *Industrial Laser Annual Handbook*, Penwell Pub., Tulsa, OK, 1986, 87-107.

34. Schuocker, D., "Theoretical Model of Reactive Gas Assisted Laser Cutting Including Dynamic Effects," *Proceedings of the SPIE – High Power Lasers and Their Industrial Applications*, Vol. 650 (1986), 210-219.

35. Sepold, G., W. Juptner, and J. Telepski, "Measuring the Quality of High Power Laser Beams," *Proceedings, SPIE Conference on High Power Lasers and Their Industrial Applications*, Vol. 605 (1986), 167-169.

36. Sliney, D.H., "Experience With Laser Safety in the USA – A Review," *Proceedings of the SPIE – High Power Lasers and Their Industrial Applications*, Vol. 650 (1986), 320-326.

37. Steen, W.M. and J.N. Kamulu, *Materials Processing: Theory and Practice, Vol. 3*, IFS, Berlin, Germany, 1983, 18-29.

38. Wallace, R., M. Bass, and S. Copley, "Curvature of Laser-Machined Grooves in Si_3N_4," *J. Applied Physics*, Vol. 59, No. 10 (May 1986), 3555-3560.

39. Ward, B.A., "Supersonic Characteristics of Nozzles Used With Lasers for Cutting," *ICALEO*, Vol. 44 (1984), 94-101.

40. Weber, H.P. and W. Hodel, "High-Power Laser Transmission Through Optical Fibers for Materials Processing," *Industrial Laser Annual Handbook*, Penwell Pub., Tulsa, OK, 1987, 33-39.

41. Weedon, T.M., "Application of Solid State Lasers in Manufacturing Industry," *Proceedings of the First International Conference on Lasers in Manufacturing* (Nov. 1983), 1-12.

4
Heat Transfer and Fluid Mechanics for Laser Machining

This chapter introduces some of the basic concepts in heat transfer, fluid mechanics and numerical solution methods. Since laser machining is a thermal process, an understanding of issues in conduction heat transfer, convection heat transfer, radiation heat transfer, and fluid mechanics is necessary in order to develop process models which relate the operating parameters of the laser, positioning system, and gas jet to the resulting erosion front geometry and temperature distribution. Also, in some situations the laser machining problem is difficult to solve analytically; in these cases, a numerical approach must be taken. Some basic concepts in finite difference and finite element methods are also introduced.

4.1 Introduction

Laser machining is a thermal process by which material is removed through a phase change, either melting or vaporization. Additionally, many of the secondary phenomena relating to surface quality which occur during laser machining, such as micro-cracking, formation of a heat-affected zone and formation of striations, can also be related to the thermal effect of the laser beam.

The need for understanding heat transfer and fluid mechanics concepts can be seen by examining the interactions which take place during laser machining (Fig. 4-1). First, the laser beam can be modelled as an radiant energy source with a prescribed spatial and temporal intensity distribution. When the laser beam impinges on the surface of the erosion front, portions of this radiant energy may be reflected back to the environment, transmitted into the workpiece interior, or absorbed by the surface; this interaction represents *radiation* heat transfer. The fraction of total beam energy absorbed at the erosion front depends on the radiation properties of the material and the erosion front geometry.

The absorbed beam energy is transformed into thermal energy through lattice vibrations in the material at the erosion front. An increase in lattice vibration results in an increase in the temperature at that location. When the local temperature reaches the melting or vaporization temperature, a phase change will occur. Some of the absorbed thermal energy may be transferred to atoms in the workpiece interior through lattice vibrations; this effect is called *conduction* heat transfer. The presence of conduction dissipates thermal energy which would otherwise be used to melt or vaporize additional material.

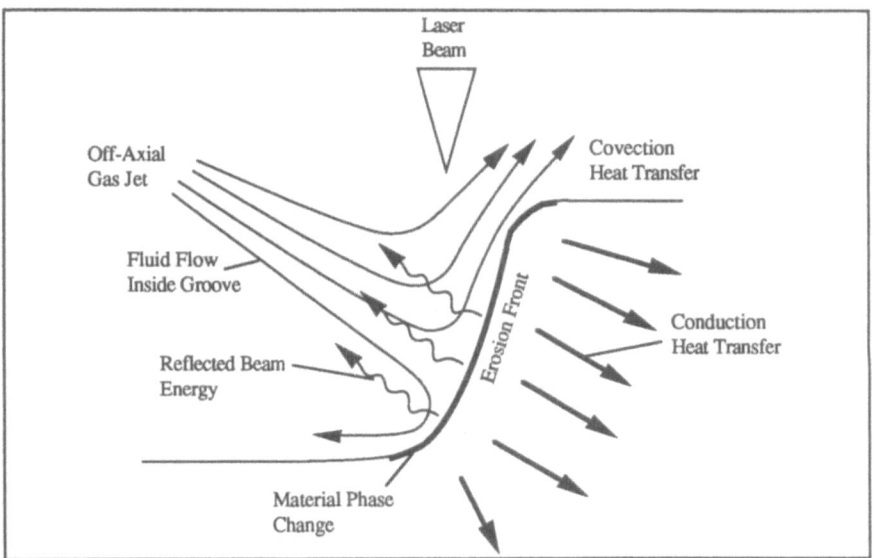

FIGURE 4.1   Example of Heat Transfer and Fluid Mechanics Occurring During Laser Machining

Thermal energy may also be dissipated from the erosion front surface to the environment under the influence of fluid flow; this interaction is called *convection* heat transfer. Since a coaxial or off-axial gas jet is usually used in tandem with the laser beam, the flow of the gas jet in the hole, kerf, or groove will influence thermal dissipation. Convection heat transfer is influenced by the *fluid mechanics* of the gas flow near the erosion front, which is related to the gas jet pressure, jet orientation with the erosion front, and the type of gas used.

4.2   Fundamentals of Heat Transfer

The analysis of the laser machining process involves a number of heat transfer phenomena occurring while the laser beam interacts with the workpiece material. There are three fundamental modes of heat transfer:

- Conduction

    A temperature gradient within a homogeneous substance results in an energy transfer rate within the medium (Fig. 4.2) which can be calculated by

$$q = -kA\frac{\partial T}{\partial n}$$

(4-1)

where $\partial T/\partial n$ is the temperature gradient in the direction normal to the area A. and q is the rate of heat transfer across area A The *thermal conductivity* k is a material property, and it may depend upon temperature and pressure. The minus sign in *Fourier's law* (Eq. 4-1) is required because energy transfer resulting from a thermal gradient must be from a region of relative high temperature to a region of relative low temperature.

The thermal conductivity of the *solid* phase of a metal of known composition is primarily dependent only upon temperature. In general, k for a pure metal decreases with temperature; alloying elements tend to reverse this trend. The thermal conductivity of a metal can usually be represented over a wide range of temperature by

$$k = k_0(1 + b\theta + c\theta^2) \tag{4-2}$$

where $\theta = T - T_{ref}$ and k_0 is the conductivity at the reference temperature T_{ref}. For materials in engineering applications, the range of temperature is relatively small, a few hundred degrees, and

$$k = k_0(1 + b\theta) \tag{4-3}$$

The thermal conductivity of a *nonhomogeneous* material is usually dependent upon the apparent bulk density, which is the mass of the substance divided by the total volume occupied. This total volume includes the void volume, such as air pockets within the overall boundaries of the piece of material. As a general rule, k for a nonhomogeneous material increases both with increasing temperature and increasing apparent bulk density.

For some *liquids* of engineering importance, k is usually temperature dependent but insensitive to pressure. Thermal conductivities of most liquids decrease with increasing temperature. The exception is water, which exhibits increasing k up to about 145°C and decreasing k thereafter. Water has the highest thermal conductivity of all liquids except the so-called liquid metals.

The thermal conductivity of a *gas* increases with increasing temperature but is essentially independent of pressure for pressures close to atmospheric. For high pressure (i.e. pressure of the order of the critical pressure or greater), the effect of pressure is significant.

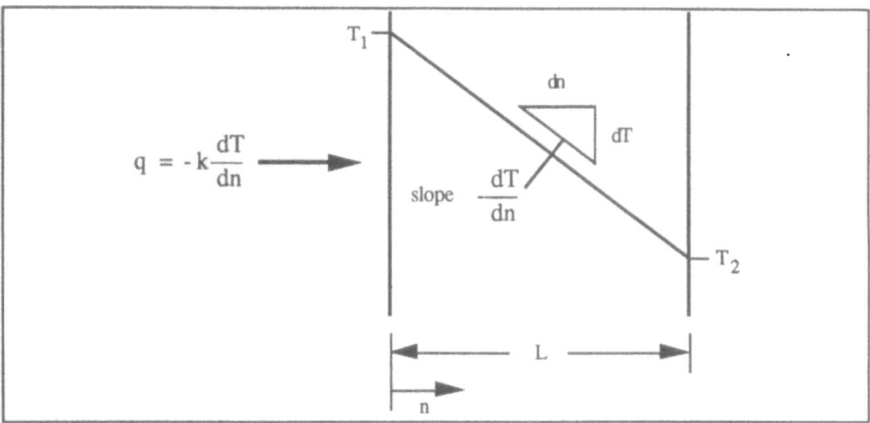

FIGURE 4.2 One-Dimensional Steady-State Conduction

Steady-state heat transfer occurs whenever the temperature at every point within the body, including the surfaces, is independent of time. If the temperature changes with time, energy is either being stored or removed from the body. This storage rate is

$$q_{stored} = mc_P \frac{\partial T}{\partial t}$$

(4-4)

where m is the mass of a body with uniform density ρ, c_p is the specific heat, and $\partial T/\partial t$ is the rate of temperature change in the body. The *specific heat* per unit mass of a substance is a measure of the variation of its stored energy with temperature. From thermodynamics the two important specific heats are:

specific heat at constant volume:
$$c_V \equiv \left(\frac{\partial u}{\partial T}\right)_V$$
(4-5)

specific heat at constant pressure:
$$c_P \equiv \left(\frac{\partial h}{\partial T}\right)_P$$
(4-6)

Here u is the integral energy per unit mass and h is the enthalpy per unit mass. In general, u and h are functions of two variables: temperature and specific volume, or temperature and pressure. For substances which are incompressible such as solids and liquids, c_p and c_v are numerically equal. For gases, however, the two specific heats may have different values.

For solids, specific heat is only weakly dependent upon temperature and even less dependent on pressure. Specific heats of liquids are influenced by temperature but not pressure. Gas specific heat values exhibit a strong temperature dependence and can be affected by pressure near the critical state.

- Convection

Whenever a solid body is exposed to a moving fluid having a temperature different from that of the body, energy is carried or *convected* away by the fluid. Convection heat transfer is actually a subset of conduction; heat is conducted from the solid surface to the fluid, which then removes the thermal energy through mass flow. If the ambient temperature of the fluid away from the surface is T_∞, the surface temperature of the solid is T_S, and the surface area exposed to the fluid A, the heat transfer per unit time is given by

$$q = hA(T_S - T_\infty) \qquad (4\text{-}7)$$

which is known as *Newton's law of cooling*. This equation defines the *convective heat-transfer coefficient h* as the constant of proportionality relating the heat transfer per unit time and unit area to the overall temperature difference. h is a function of flow geometry, flow velocity and fluid properties.

- Radiation

The third mode of heat transfer is due to electromagnetic wave propagation, which can occur in a total vacuum as well as in a medium. Experimental evidence indicates that radiant heat transfer is proportional to the fourth power of the absolute temperature, whereas conduction and convection are proportional to a linear temperature difference. In some cases, an idealized surface can be assumed to absorb all of the incoming radiation; in this case, the body of the surface is termed a *black body* and the radiation heat transfer from this body follows the *Stefan-Boltzmann law*.

$$q = \sigma A T^4 \qquad (4\text{-}8)$$

where T is the absolute temperature, A the surface area and σ is the Stefan-Boltzmann constant, which is independent of surface, medium, and temperature; its value is $5.6697 \times 10^{-8} W/m^2\text{-}k^4$.

A real surface, however, absorbs less energy than a black body and is modelled as a *gray body* by specifying an emissivity value ε, which ranges from zero to one. Additionally, if wave propagation occurs through a gas or vapor, the medium may absorb a portion of the thermal radiation. In this case, the medium can be modelled as a *gray gas*. The radiation heat transfer for gray bodies and gases follow the relationship:

$$q = \varepsilon \sigma A T^4 \qquad (4\text{-}9)$$

4.3   Conduction

4.3.1   General Conduction Heat Equation

The objective of a conduction heat transfer analysis is to estimate the temperature distribution and heat transfer rate within a solid, liquid or gas with no relative motion within the body. From Eq. (4-1), the heat transfer rate can be related to the local temperature gradient. In some cases the temperature gradient can be easily derived through physical measurements of temperature at predetermined locations within the body. However, in more complex cases such as the ones occurring during laser machining, the temperature gradients can only be established by considering an energy balance equation, which determines the temperature distribution within a region. From the temperature distribution, the temperature gradient at any desired location within the material can be formed and consequently the heat transfer rate may be calculated.

Consider a control volume consisting of a small parallelepiped, as shown in Figure 4.3. This may be an element of material from a homogeneous solid or a homogeneous fluid so long as there is no relative motion between the macroscopic material particles. Heating of the material results in an energy flux per unit area within the control volume. This flux is, in general, a three-dimensional vector. For simplicity, only one component, q_x, is shown in Figure 4.3. For the control volume in Figure 4.3, and for an incompressible substance, the net work done on the control volume is converted to internal energy, u, by applying the first law of thermodynamics. The rate of energy conversion from work, chemical reaction, etc. can be denoted as q^m:

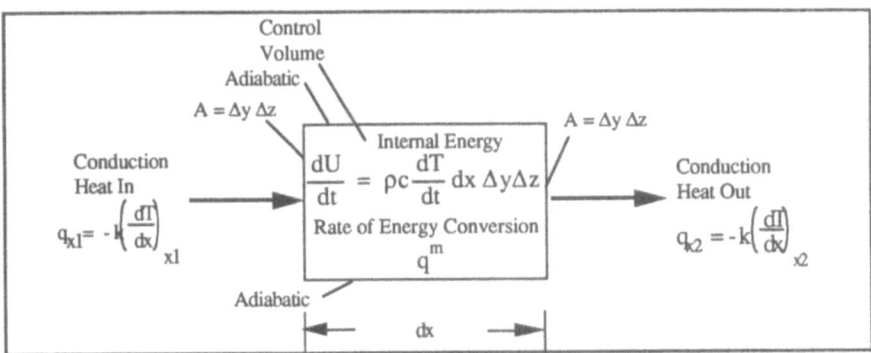

FIGURE 4.3   Control Volume for One-Dimensional Time-Dependent Conduction

$$q_{x_1} + q_{y_1} + q_{z_1} + q^m \Delta x \Delta y \Delta z = q_{x_2} + q_{y_2} + q_{z_2} + \frac{\partial U}{\partial t}$$

(4-10)

In the x-direction the two terms may be grouped to form

$$q_{x_1} - q_{x_2} = -\Delta y \Delta z \left[\left(k \frac{\partial T}{\partial x} \right)_{x_1} - \left(k \frac{\partial T}{\partial x} \right)_{x_2} \right]$$

(4-11)

by application of Fourier's law. k may be temperature dependent and therefore spatially dependent. By a Taylor's series expansion,

$$\left(k \frac{\partial T}{\partial x} \right)_{x_1} = k \frac{\partial T}{\partial x} + \left(\frac{\Delta x}{2} \right) \frac{\partial}{\partial x} \left(k \frac{\partial T}{\partial x} \right) + \dots$$

(4-12a)

$$\left(k \frac{\partial T}{\partial x} \right)_{x_2} = k \frac{\partial T}{\partial x} + \left(\frac{\Delta x}{2} \right) \frac{\partial}{\partial x} \left(k \frac{\partial T}{\partial x} \right) + \dots$$

(4-12b)

so that Eq. (4-11) becomes

$$q_{x_1} - q_{x_2} = \Delta y \, \Delta z \left[\Delta x \frac{\partial}{\partial x} \left(k \frac{\partial T}{\partial x} \right) + \dots \right]$$

(4-13)

Similarly for the y and z directions,

$$q_{y_1} - q_{y_2} = \Delta x \, \Delta z \left[\Delta y \frac{\partial}{\partial y} \left(k \frac{\partial T}{\partial y} \right) + \dots \right]$$

(4-14)

$$q_{z_1} - q_{z_2} = \Delta x \, \Delta y \left[\Delta z \frac{\partial}{\partial z} \left(k \frac{\partial T}{\partial z} \right) + \dots \right]$$

(4-15)

Finally, the internal energy storage per unit volume and per unit temperature is the product of density and specific heat, so

$$\frac{\partial U}{\partial t} = \rho c \left(\Delta x \, \Delta y \, \Delta z \right) \frac{\partial T}{\partial t}$$

(4-16)

Substituting Eqs. 4.13 through 4.16 into 4.10, dividing by the volume $\Delta x \, \Delta y \, \Delta z$, and taking the limit as Δx, Δy, and Δz simultaneously approach zero yields the *general conduction equation*:

$$\frac{\partial}{\partial x} \left(k \frac{\partial T}{\partial x} \right) + \frac{\partial}{\partial y} \left(k \frac{\partial T}{\partial y} \right) + \frac{\partial}{\partial z} \left(k \frac{\partial T}{\partial z} \right) + q''' = \rho c \frac{\partial T}{\partial t}$$

(4-17)

for the temperature T as a function of x, y, z and t. Here, k is the thermal conductivity, ρ is the density, c is the specific heat per unit mass, and q" is the *rate of internal energy conversion ("heat generation")* per unit volume.

In most engineering problems k can be taken as a constant, and Eq. (4-17) reduces to

$$\frac{\partial^2 T}{\partial x^2} + \frac{\partial^2 T}{\partial y^2} + \frac{\partial^2 T}{\partial z^2} + \frac{q'''}{k} = \frac{1}{\alpha}\frac{\partial T}{\partial t}$$

(4-18)

where α, defined as $\alpha = k/(\rho c_p)$, is the thermal diffusivity. Thermal energy diffuses rapidly through substances with high α and slowly through those with low α.

If the thermal properties depend on the temperature, the conduction equation becomes non-linear. In the case of thermal properties varying with temperature but independent of position, Eq. (4-17) becomes

$$\rho c \frac{\partial T}{\partial t} = k \nabla^2 T + q''' + \frac{\partial k}{\partial T}\left(\left(\frac{\partial T}{\partial x}\right)^2 + \left(\frac{\partial T}{\partial y}\right)^2 + \left(\frac{\partial T}{\partial z}\right)^2\right)$$

(4-19)

In this case, Eq. (4-19) can be simplified by introducing a variable Θ which lumps both conductivity and temperature.

$$\Theta = \frac{1}{k_o}\int_0^T k\, dT$$

(4-20)

where k_o is the value of k when $T = T_0$, a reference temperature. Eq. (4-19) becomes

$$\nabla^2\Theta - \frac{1}{\alpha}\frac{\partial\Theta}{\partial t} = -\frac{q'''}{k_0}$$

(4-21)

In Eq. (4-21) α is expressed as a function of the variable Θ, so the general form of Eq. (4-18) is preserved, but with a diffusivity α which depends on Θ. In many cases the variation of α with temperature is much less than the variation of k with temperature, so that, to a reasonable approximation α may be taken to be constant.

4.3.2   Theory of a Moving Heat Source

In the particular case of laser machining, the beam can be viewed generally as a heat source which impinges on the surface of the workpiece. If the beam is stationary such as in the case of laser drilling, then the erosion front moves relative to the laser beam. This results in a temperature distribution which changes with time; therefore, laser drilling is a non-steady process. In laser cutting and scribing, the laser beam is in relative motion with the workpiece; in case of constant scanning velocity the erosion front and the resulting temperature distribution is constant relative to a coordinate system fixed at the laser beam. A steady temperature field simplifies the heat transfer problem to steady-state conduction. Eq. (4-18) can be used without internal heat generation. If a constant scanning or cutting velocity is assumed, then the time dependence of temperature can be transformed into a spatial dependence.

$$\frac{\partial T}{\partial t} = \frac{\partial T}{\partial x}\frac{dx}{dt} = -v\frac{\partial T}{\partial x}$$

(4-22)

and Eq. (4-18) becomes

$$\frac{\partial^2 T}{\partial x^2} + \frac{\partial^2 T}{\partial y^2} + \frac{\partial^2 T}{\partial z^2} = -\frac{v}{\alpha}\frac{\partial T}{\partial x}$$

(4-23)

Rosenthal [13] transformed the conduction equation (4-23) into a simpler form by assuming a temperature distribution of the form:

$$T = T_o + e^{-vx/2\alpha}\phi(x,y,z)$$

(4-24)

where T_o is the initial temperature of the solid and ϕ is an unknown function of x, y, and z. Substitution of equation (4-24) into (4-23) yields

$$\frac{\partial^2 \phi}{\partial x^2} + \frac{\partial^2 \phi}{\partial y^2} + \frac{\partial^2 \phi}{\partial z^2} - \left(\frac{v}{2\alpha}\right)^2 \phi = 0$$

(4-25a)

or

$$\nabla^2 \phi - \left(\frac{v}{2\alpha}\right)^2 \phi = 0$$

$$(4\text{-}25b)$$

By using an assumed temperature profile, the general conduction equation can be reduced to a relatively simple relationship used for finding temperature distributions in infinite and semi-infinite solids.

For one-dimensional conduction, $\partial T/\partial y = \partial T/\partial z = 0$, Eq. (4-25a) is reduced to

$$\frac{d^2\phi}{dx^2} - \left(\frac{v}{2\alpha}\right)^2 \phi = 0$$

$$(4\text{-}26)$$

The boundary conditions are

$$\frac{dT}{dx} \to 0 \ \text{ for } x \to \infty$$

$$(4\text{-}27)$$

Furthermore, considering a plane source with a rate of heat q" per unit area

$$k\left(\frac{dT}{dx}\right)_{x \to 0^-} + k\left(\frac{dT}{dx}\right)_{x \to 0^+} = q"$$

$$(4\text{-}28)$$

A general solution of Eq. (4-25) can be written as

$$\phi = C_1 \, e^{vx/2\alpha} + C_2 \, e^{-vx/2\alpha}$$

$$(4\text{-}29)$$

where C_1 and C_2 are constants to be determined.

Thus,

$$T - T_o = C_1 \, e^{-vx/\alpha} + C_2$$

$$(4\text{-}30)$$

From (4-27) $C_1 = 0$ when x<0; therefore for x<0:

$$T - T_o = C_2$$

$$(4\text{-}31)$$

Eq. (4-31) shows that none of the heat produced in the plane leaves in the negative x direction, i.e.

$$\left(\frac{dT}{dx}\right)_{x \to 0^-} = 0$$

(4-32a)

Hence, all of the heat must be conducted away in the positive x direction, or

$$-k\left(\frac{dT}{dx}\right)_{x \to 0^+} = q''$$

(4-32b)

Likewise $C_2 = 0$, when x>0, therefore for x>0:

$$T - T_o = C_1 e^{-vx/\alpha}$$

(4-33)

Also, since $T - T_0$ must have the same value at x=0, C_1 must equal C_2. On the other hand, with reference to equation (4-28)

$$C_1(vxk/\alpha) = q''$$

(4-34)

or

$$C_1 = C_2 = q''/c_p\rho v$$

(4-35)

thus for x<0

$$T - T_o = \frac{q''}{\rho c_p v}$$

(4-36)

and for x>0

$$T - T_o = \frac{q''}{c_p\rho v} e^{-vx/\alpha}$$

(4-37)

From equation (4-37) it follows that $q''/c_p\rho v$ is the rise of temperature at the location of the source, i.e., for x = 0, and corresponds to the maximum value of temperature reached in the solid (Fig. 4.4).

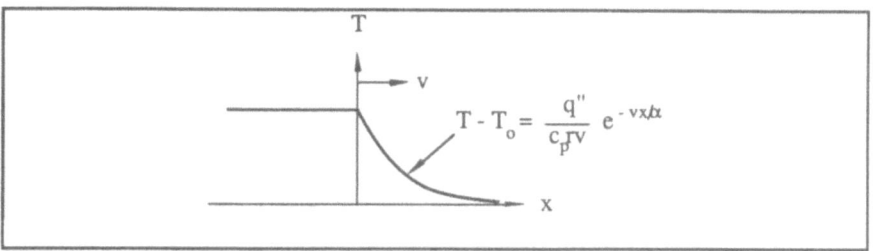

If we assume that heat flows in the x- and y-direction, and no flow in the z-direction
(two-dimensional case) then $\partial T/\partial z = 0$, and equation (4-24) becomes

$$\frac{\partial^2 \phi}{\partial x^2} + \frac{\partial^2 \phi}{\partial y^2} = \left(\frac{v}{2\alpha}\right)^2 \phi$$

(4-38)

The boundary conditions are

$$\frac{\partial T}{\partial x} \to 0 \ \ as \ x \to \pm\infty$$

$$\frac{\partial T}{\partial y} \to 0 \ \ as \ y \to \pm\infty$$

(4-39)

and, by considering a circle $2\pi r$ drawn around the heat source with radius
$r = (x^2 + y^2)^{1/2}$

$$- 2\pi r k \frac{\partial T}{\partial r} \to q' \ \ for \ r \to 0$$

(4-40)

where q' is the rate of heat in a linear source.

ϕ is symmetric with respect to r, although T does not show radial symmetry (Fig.
4.5). Because of the boundary conditions and the symmetrical form of Eq. (4-38) with
respect to x and y, ϕ depends only on the distance from the heat source, the equation (4-
38) becomes in cylindrical coordinates

$$\frac{d^2\phi}{dr^2} + \frac{1}{r}\frac{d\phi}{dr} - \left(\frac{v}{2\alpha}\right)^2 \phi = 0$$

(4-41)

The solution of Eq. (4-41) satisfying the boundary condition Eq. (4-39) is known as the modified Bessel function of the second kind and zero order, and is represented by the symbol $K_0(vr/2\alpha)$. It can be shown that this function approaches ln (r) as r approaches zero, and (dK_0/dr) approaches a constant value, which means that $K_0(vr/2\alpha)$ fulfills the condition Eq. (4-40). On the other hand $(\pi/\alpha vr)^{1/2} e^{-vr/2\alpha}$ approaches zero as r tends to infinity, whereby the boundary condition Eq. (4-39) is also satisfied. Therefore, the solution of the two-dimensional case (Fig. 4.5) is

$$T - T_0 = \frac{q'}{2\pi k} e^{-vx/2\alpha} K_0\left(\frac{vr}{2\alpha}\right)$$

(4-42)

For the three-dimensional case (heat conduction in x, y, and z), the boundary conditions are

$$\frac{\partial T}{\partial x} \to 0 \quad \text{as } x \to \pm\infty$$

$$\frac{\partial T}{\partial y} \to 0 \quad \text{as } y \to \pm\infty$$

$$\frac{\partial T}{\partial z} \to 0 \quad \text{as } z \to \pm\infty$$

(4-43)

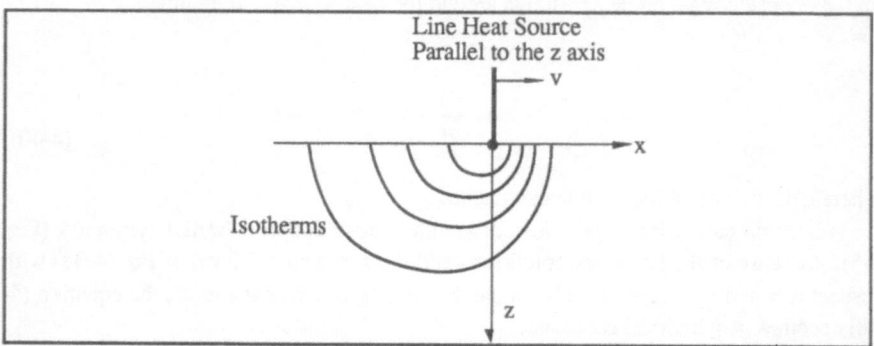

FIGURE 4.5   Temperature Distribution in a Semi-Infinite Body Due to Heat Source Moving in x-Direction

and by considering a spherical surface $4\pi R^2$ drawn around the heat source with radius $R = (x^2 + y^2 + z^2)^{1/2}$

$$- 4\pi R^2 k \frac{\partial T}{\partial R} \to q \text{ for } R \to 0 \qquad (4\text{-}44)$$

where q is the rate of heat of a point source.

Because of conditions (4-43) and (4-44), and the symmetrical form of the differential equation (4-24), ϕ depends only on the distance R, thus using polar coordinates equation (4-24) can be rewritten as follows

$$\frac{d^2\phi}{dR^2} + \frac{2}{R} \frac{d\phi}{dR} - \left(\frac{v}{2\alpha}\right)^2 \phi = 0 \qquad (4\text{-}45)$$

which results in the expression

$$\frac{d^2\phi}{dR^2} + \frac{2}{R} \frac{d\phi}{dR} = \frac{1}{R} \frac{d^2(R\phi)}{dR^2} \qquad (4\text{-}46)$$

thus equation (4-45) becomes

$$\frac{d^2(R\phi)}{dR^2} - \left(\frac{v}{2\alpha}\right)^2 R\phi = 0 \qquad (4\text{-}47)$$

A solution of equation (4-46) satisfying the boundary conditions (4-43) is

$$R\phi = C\, e^{-\, vR/2\alpha} \qquad (4\text{-}48a)$$

where C is a constant,
or

$$\phi = \frac{C}{R} e^{-\, vR/2\alpha} \qquad (4\text{-}48b)$$

This solution also satisfies condition (4-44) since $(d\phi/dR)R^2$ tends to a constant value as R approaches to zero. The corresponding temperature distribution can be determined as

$$T - T_o = \frac{q}{4\pi k} e^{-vx/2\alpha} \frac{e^{-vR/2\alpha}}{R}$$

$$(4\text{-}49)$$

4.4   Convection and Fluid Mechanics

The study of convective heat transfer includes the behavior of fluids in motion, since thermal energy is transported through fluid flow. Therefore the discussion of the fundamental relationships in convective heat transfer will be preceded by the introduction of some basic relationships from fluid mechanics.

4.4.1   Hydrostatics

The pressure differential, $p_2 - p_1$, between two points in a static or uniformly moving fluid is proportional to the difference in elevation, $y_2 - y_1$, between the points, the fluid density ρ, and the local acceleration of gravity g.

$$p_2 - p_1 = -\rho g \, (y_2 - y_1)$$

$$(4\text{-}50)$$

If elevation y_1 is taken as a datum at zero pressure, Eq. (4-50) simplifies to

$$p = \gamma h$$

$$(4\text{-}51)$$

where $\gamma = \rho g$; Eq. (4-51) is known as the *hydrostatic equation*.

A body undergoes an apparent loss of weight when partially or wholly immersed in a fluid mainly due to the *buoyant force*, F_B, which is the vertical resultant of the pressure distribution exerted by the fluid on the body. *Buoyancy*, which is central in natural convection, is governed by *Archimedes' principle. The buoyant force is equal to the weight of fluid displaced.* The line of action of the buoyant force is through the centroid of the volume of fluid displaced.

4.4.2   Fluid Dynamics

A streamline is an imaginary line, taken at an instant of time in a flow field, such that the fluid velocity at every point of the line is tangent to it. Since movement occurs only in the direction of the velocity vector, no mass crosses a streamline.

A *stream tube* is a finite surface, made up of an infinite number of streamlines, across which there is no flow.

By noting that the velocity components in the x- and y-directions are

$$u = \frac{dx}{dt} \quad v = \frac{dy}{dt}$$

$$(4\text{-}52)$$

The differential equations of a streamline can be derived by eliminating dt, giving

$$u \, dy = v \, dx \qquad (4\text{-}53)$$

Similarly, for the three-dimensional case

$$v \, dz = w \, dy \qquad (4\text{-}54)$$
$$w \, dx = u \, dz \qquad (4\text{-}55)$$

If, at the specified instant, u, v, and w are known functions of position, any two of Eq.(4.53) through (4-55) may be integrated to give the equation of the streamlines.

When concentrating attention on a fixed region in space without regard to the identities of the fluid particles within it at a given time – known as the Eulerian approach, as contrasted with the Lagrangian method, which focuses on the motion of individual particles – the velocity field is given in cartesian coordinates by

$$\mathbf{V} = \mathbf{i}u + \mathbf{j}v + \mathbf{k}w \qquad (4\text{-}56)$$

where the velocity components are functions of space and time, i.e.

$$u = u\,(x,\, y,\, z,\, t) \qquad (4\text{-}57)$$
$$v = v\,(x,\, y,\, z,\, t)$$
$$w = w\,(x,\, y,\, z,\, t)$$

Using the chain rule for partial differentiation, the rate of change of the velocity V is given by

$$\mathbf{a} = \frac{d\mathbf{V}}{dt} = \frac{\partial \mathbf{V}}{\partial x}\frac{dx}{dt} + \frac{\partial \mathbf{V}}{\partial y}\frac{dy}{dt} + \frac{\partial \mathbf{V}}{\partial z}\frac{dz}{dt} + \frac{\partial \mathbf{V}}{\partial t} \qquad (4\text{-}58)$$

Since, for a moving particle, $u = dx/dt$, and $w = dz/dt$, Eq. (4-58) may be written

$$\mathbf{a} = \frac{d\mathbf{V}}{dt} = \left(u\frac{\partial \mathbf{V}}{\partial x} + v\frac{\partial \mathbf{V}}{\partial y} + w\frac{\partial \mathbf{V}}{\partial z}\right) + \frac{\partial \mathbf{V}}{\partial t} \qquad (4\text{-}59)$$

which is known as the substantial, total, or fluid derivative, designated DV/Dt. The influence of time on a particle's behavior is given by the local acceleration, $\partial \mathbf{V}/\partial t$; space dependence is given by the convective acceleration, the terms in parentheses.

When the local acceleration is zero, $\partial \mathbf{V}/\partial t = 0$, the motion is *steady*. Even though the velocity may change with respect to space, it does not change with respect to time. Streamlines are fixed in steady flow. A flow which is time dependent, $\partial \mathbf{V}/\partial t \neq 0$, is *unsteady*.

Uniform flow occurs when the convective acceleration is zero. The velocity vector is identical at every point in the flow field. The flow may be unsteady, but the velocity must change identically at every point. Streamlines are straight. An example is a frictionless fluid flowing through a long straight pipe. *Nonuniform* flow is space dependent. A frictionless fluid would flow nonuniformly in a pipe elbow.

In *laminar* flow, fluid particles move very smoothly parallel to each other. A dye stream injected in a laminar flow field would move in a thin line. Low velocities in smooth channels can produce laminar flow. At high velocities, however, *turbulent* flow, characterized by random motion of fluid particles, occurs. A dye injected in the stream would break up and diffuse throughout the flow field. Turbulent flow is always unsteady in the strict sense.

The type of flow depends upon a comparison of the viscous and inertial components of fluid flow. A nondimensional parameter, called the *Reynolds number*, is defined as

$$Re \equiv \frac{Vl}{v} \tag{4-60}$$

where l is the characteristic length, V is the average fluid velocity, $v \equiv \mu_m/\rho$ is the kinematic viscosity and μ_m the viscosity of the flowing fluid. For flow over a flat plate, the distance from the leading edge of the plate is the characteristic length, whereas for pipe flow, the pipe diameter D is the characteristic length. In flow over a flat plate, the transition from laminar to turbulent flow commonly occurs at $300,000 < Re < 600,000$, while in pipe flow turbulence occurs for $Re > 2000$.

4.4.3 Conservation of Mass and Equation of Motion

For steady flow the mass entering a stream tube is equal to mass leaving. Thus, if ṁ denotes the rate of mass transport through a cross section,

$$\dot{m} = \rho_1 A_1 V_1 = \rho_2 A_2 V_2 = \text{constant} \tag{4-61}$$

where V is the average velocity taken normal to the cross-sectional area A, and ρ is the density, assumed uniform over a cross section. This equation is known as the *continuity equation*. The *mass velocity*, $G \equiv \rho V$, is often used in heat transfer calculations, giving

$$\dot{m} = AG = \text{constant} \tag{4-62}$$

If, in addition to being steady, the flow is *incompressible* (ρ = constant), the continuity equation reduces to

$$Q = A_1 V_1 = A_2 V_2 = \text{constant} \tag{4-63}$$

where Q is the *volumetric flow rate*.

The differential form of the continuity equation, which holds for steady or unsteady flow, is in cartesian coordinates,

$$\frac{\partial \rho}{\partial t} + \frac{\partial(\rho u)}{\partial x} + \frac{\partial(\rho v)}{\partial y} + \frac{\partial(\rho w)}{\partial z} = 0 \tag{4-64}$$

and in general vector form,

$$\frac{\partial \rho}{\partial t} + \nabla \cdot (\rho V) = 0$$

$$(4\text{-}65)$$

where $\nabla \cdot$ is the *divergence*, which may be conveniently expressed in any orthogonal coordinate system. If the flow is steady and incompressible, the continuity equation in the cartesian coordinate system reduces to

$$\nabla \cdot V = \frac{\partial u}{\partial x} + \frac{\partial v}{\partial y} + \frac{\partial w}{\partial z} = 0$$

$$(4\text{-}66)$$

Using Newton's second law, one can derive an equation for the motion of any fluid along a streamline:

$$\frac{1}{\rho}\frac{\partial p}{\partial s} + g\frac{\partial z}{\partial s} + \frac{\tau}{\rho R_h} + \frac{\partial}{\partial s}\left(\frac{V^2}{2}\right) + \frac{\partial V}{\partial t} = 0$$

$$(4\text{-}67)$$

Here, p is the pressure along the streamline, s is the arc length along the streamline, z is the vertical coordinate, τ is the surface shear stress and R_h is the hydraulic radius. Eq. (4-67) is valid for viscous or frictionless fluid and for steady or unsteady flow.

Assuming a uniform gravitational field (g = constant) and steady, incompressible flow ($\partial V/\partial t = 0$, ρ = constant), Eq. (4-67) can be integrated along the streamline from s = s_1 to s = s_2. The result is:

$$\frac{p_2 - p_1}{\rho} + g(z_2 - z_1) + \frac{V_2^2 - V_1^2}{2} + h_L = 0$$

$$(4\text{-}68)$$

where h_L is the head loss due to friction.
For steady, incompressible flow, whereby the frictional losses are expressed as a head loss:

$$(\Delta p)_{loss} = \rho g h_L$$

$$(4\text{-}69)$$

If the flow is also frictionless, ($h_L = 0$) Eq. (4-68) reduces to Bernoulli's equation:

$$\frac{p_1}{\gamma} + z_1 + \frac{V_1^2}{2g} = \frac{p_2}{\gamma} + z_2 + \frac{V_2^2}{2g} = \text{constant}$$

$$(4\text{-}70)$$

where $\gamma = \rho g$. The "constant" in Eq. (4-70) will, in general, vary from streamline to streamline.

The terms of Bernoulli's equation represent mechanical energy possessed by the flowing fluid due to pressure, position and velocity. If, in addition to these energy terms, energy is added to or extracted from a given flowing mass, an energy balance requires that

$$\left(\frac{p_1}{\rho} + gz_1 + \frac{V_1^2}{2}\right) + q - gh_L - w_s = \left(\frac{p_2}{\rho} + gz_2 + \frac{V_2^2}{2}\right) \tag{4-71}$$

where q is the heat transfer, positive when added to the system, and w_s is the work transfer, positive when done by the system. Each term in Eq. (4-71) is an energy per unit mass.

Identifying the frictional loss with the gain in internal energy between Stations 1 and 2, i.e. $gh_L = u_2 - u_1$, puts Eq. (4-71) into the more convenient form

$$\left(\frac{p_1}{\rho} + u_1 + gz_1 + \frac{V_1^2}{2}\right) + q - w_s = \left(\frac{p_2}{\rho} + u_2 + gz_2 + \frac{V_2^2}{2}\right) \tag{4-72}$$

In order to analyze the velocity profile of a fluid, an infinitesimal control volume can be taken (Fig. 4.6). The forces which can act on this control volume include the force from a pressure gradient, the viscous shear force from the surrounding fluid, and the inertial acceleration of the fluid in the control volume. Taking a force balance for the control volume in the x-direction results in the equation:

$$p \, dy - \left(p + \frac{dp}{dx}dx\right)dy + \left(\tau_{yx} + \frac{\partial \tau_{yx}}{\partial y}\right)dx - \tau_{yx}dx =$$
$$\left(\rho u^2 + \frac{\partial \rho u^2}{\partial x}\right)dy - \rho u^2 \, dy + \left(\rho uv + \frac{\partial \rho uv}{\partial y}dy\right)dx - \rho uv \, dx \tag{4-73}$$

Eq. (4-73) can be simplified to:

$$\frac{\partial \tau_{yx}}{\partial y}dydx - \frac{dP}{dx}dxdy = \left(\frac{\partial \rho u^2}{\partial x} + \frac{\partial \rho uv}{\partial y}\right)dxdy \tag{4-74}$$

The simple equation for shear stress on a surface is:

$$\tau_{laminar} = \mu_f \frac{\partial u}{\partial n} \tag{4-75}$$

where u is the velocity parallel to the surface and n is the direction normal to the surface.

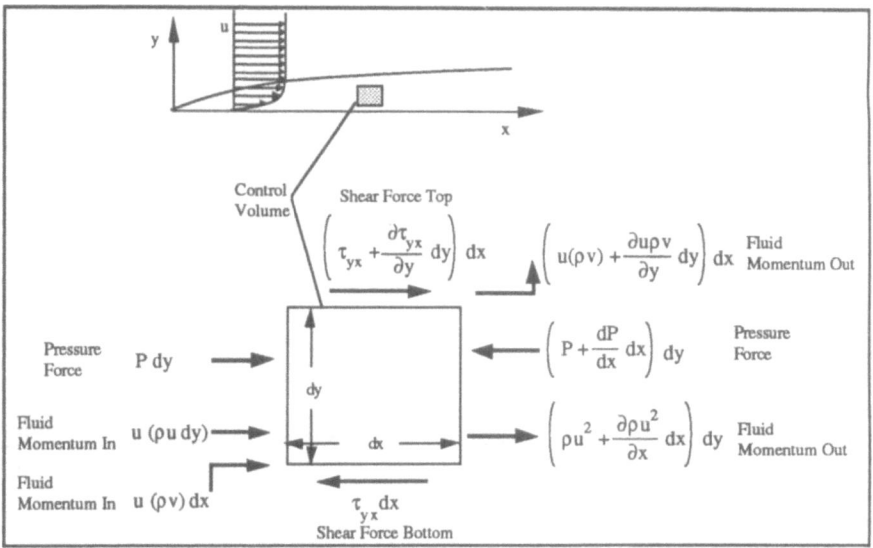

FIGURE 4.6   Fluid Control Volume

In turbulent flow, the effective shear stress is a function of both fluid viscosity and the eddy viscosity ε, which is related to the fluid motion.

$$\tau_{turb} = (\mu_f + \rho\varepsilon)\frac{du}{dn}$$

(4-76)

Assuming laminar flow, Eq. (4-75) can be substituted into Eq. (4-74) to form the *Navier-Stokes equation* in the x direction:

$$u\frac{\partial u}{\partial x} + v\frac{\partial u}{\partial y} = -\frac{1}{\rho}\frac{dP}{dx} + v\frac{\partial^2 u}{\partial y^2}$$

(4-77)

4.4.4   Boundary Layer Flow

Heat transfer by convection is normally controlled within a thin layer of the fluid, adjacent to the immersed body. The quantity of heat transferred depends upon the fluid motion and the viscous effects within this boundary layer. If fluid properties do not vary with temperature, the boundary layer can be treated independently of the heat transfer; otherwise the heat transfer and the fluid flow processes are interwoven.

A fluid moves either due to an enforced pressure drop or due to buoyancy effects (density differences). Pressure-drop induced velocities are much larger than velocities obtained due to buoyancy; consequently the heat fluxes due to forced velocities (forced

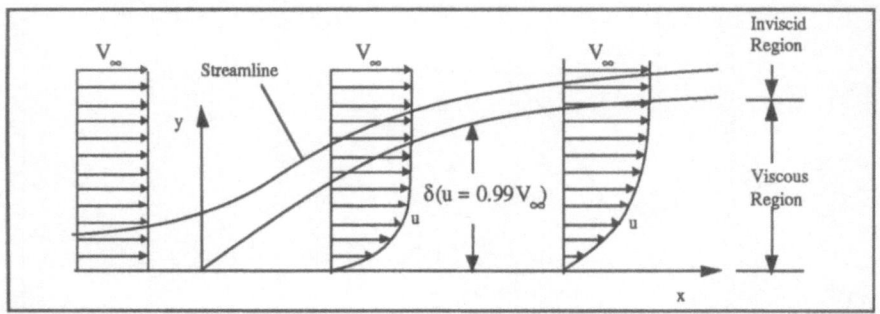

FIGURE 4.7   Incompressible Fluid Flow Over a Flat Plate

convection) are much greater than those due to buoyancy effects (free convection). Fundamentals of boundary layer flow can be understood by laminar flow along a flat plate, since the fluid motion can be visualized and since an exact solution exists for the fluid's behavior.

Consider a very thin, flat plate with an unbounded, incompressible, viscous fluid flowing parallel to it with a free-stream velocity V_∞ (Fig. 4.7). The velocity profiles at two stations are also shown; the velocity varies from zero at the surface (no-slip condition) to 0.99 V_∞ at the edge of the boundary layer. Outside the boundary layer is the inviscid region, where Bernoulli's equation is valid. Continuity requires that streamlines diverge as the fluid is retarded more and more in moving along the plate. The divergence of streamlines suggests motion in the y-direction, normal to the plate.

By applying Newton's second law and the continuity equation to an infinitesimal, two-dimensional control volume within the boundary layer, and assuming that

1. Fluid viscosity is constant
2. Shear in the y-direction is negligible
3. The flow is steady, and the fluid is incompressible

the following is derived:

x-momentum:
$$u\frac{\partial u}{\partial x} + v\frac{\partial u}{\partial y} = \frac{1}{\rho}\frac{dp}{dx} + v\frac{\partial^2 u}{\partial y^2}$$
(4-78)

y-momentum:
$$\frac{\partial p}{\partial y} = 0$$
(4-79)

continuity:
$$\frac{\partial u}{\partial x} + \frac{\partial v}{\partial y} = 0$$
(4-80)

Since the pressure can be determined in the inviscid region from Bernoulli's equation,

$$\frac{p}{\rho} + \frac{V^2}{2} = \text{constant}$$
(4-81)

the pressure at a given lengthwise location within the boundary layer is known from Eq. (4-81). Equations (4-78) and (4-81) can then be solved simultaneously to give the velocity distribution. [5]

By use of an order-of-magnitude analysis one can find a functional relationship for the boundary layer thickness δ, assuming constant fluid properties and zero pressure gradient.

The governing equations are:

continuity:
$$\frac{\partial u}{\partial x} + \frac{\partial v}{\partial y} = 0$$
(4-82)

x-momentum:
$$u \frac{\partial u}{\partial x} + v \frac{\partial u}{\partial y} = \nu \frac{\partial^2 u}{\partial y^2}$$
(4-83)

Except very close to the surface, the velocity u within the boundary layer is of the order of the free-stream velocity V_∞, i.e. $u \approx V_\infty$. And the y-dimension within the boundary layer is of the order of the boundary layer thickness, $y \approx \delta$. The continuity equation can then be approximated as

$$\frac{V_\infty}{x} + \frac{v}{\delta} \approx 0$$
(4-84)

giving

$$v \approx \frac{V_\infty \delta}{x}$$
(4-85)

Using the estimates of u, y and v in the x-momentum equation gives

$$V_\infty \frac{V_\infty}{x} + \frac{V_\infty \delta}{x} \frac{V_\infty}{\delta} \approx \nu \frac{V_\infty}{\delta^2} \qquad \delta^2 \approx \frac{\nu x}{V_\infty}$$
$$\text{or}$$
(4-86)

Eq. (4-86) can be divided by x^2 to make dimensionless.

$$\frac{\delta}{x} \approx \sqrt{\frac{\nu}{V_\infty x}} = \sqrt{\frac{1}{Re_x}}$$
(4-87)

If the velocity profiles at different locations along the plate can be assumed geometrically similar [5], then the relationship between δ and Reynolds number becomes

$$\frac{\delta}{x} \approx \frac{5.0}{\sqrt{Re_x}}$$

(4-88)

4.4.5 Boundary Layer Flow With Heat Transfer

4.4.5.1 Thermal Boundary Layer

When a fluid at one temperature flows along a surface which is at another temperature, the behavior of the fluid cannot be described by the hydrodynamic equations alone. In addition to the hydrodynamic boundary layer, a thermal boundary layer develops due to the temperature distribution.

Figure 4.8 shows temperature distributions within the thermal boundary layer, having a gradient which is infinite at the leading edge and approaches zero as the layer develops downstream. Shown also is a heat balance at the plate's surface, where the heat conducted from the plate must equal the heat convected into the fluid; thus,

$$-k\frac{\partial T}{\partial y}\Big|_s = h_x(T_s - T_\infty)$$

(4-89)

or

$$h_x = -\frac{k}{T_s - T_\infty}\left(\frac{\partial T}{\partial y}\right)_s$$

(4-90)

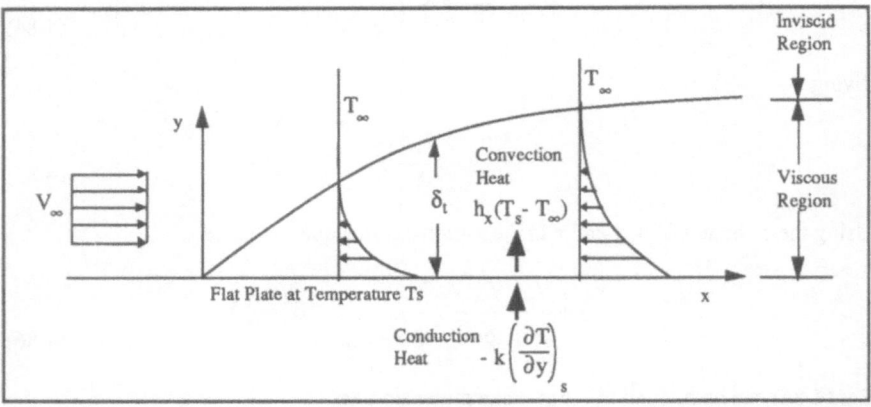

FIGURE 4.8   Temperature Distribution and Heat Balance Within the Thermal Boundary Layer ($\nu = \alpha$)

In order to obtain an expression for the convective heat-transfer coefficient h_x, one must find the temperature distribution. With the aid of an energy balance on an infinitesimal control volume within the boundary layer, the following relationship can be derived:

$$u \frac{\partial T}{\partial x} + v \frac{\partial T}{\partial y} = \alpha \frac{\partial^2 T}{\partial y^2}$$

(4-91)

where α is the thermal diffusivity
With the boundary conditions,

at $y = 0$:   (1) $T = T_s$;   (2) $\dfrac{\partial^2 T}{\partial y^2} = 0$ (4-92)

at $y = \delta_t$:   (3) $T = T_\infty$;   (4) $\dfrac{\partial T}{\partial y} = 0$

The thermal boundary layer thickness δ_t is defined as the distance required for the temperature T to reach 99 percent of its free-stream value T_∞. The assumptions for Eq. (4-91) are:

1. Steady, incompressible flow
2. Constant fluid properties evaluated at the film temperature,

$$T_f \equiv \frac{T_s + T_\infty}{2}$$

(4-93)

3. Negligible body forces, viscous heating (low velocity), and conduction in the flow direction

Recalling that the x-momentum equation for a constant-pressure field is

$$u \frac{\partial u}{\partial x} + v \frac{\partial u}{\partial y} = v \frac{\partial^2 u}{\partial y^2}$$

(4-94)

Eq. (4-94) is similar in form to the energy equation (4-18). The temperature and velocity variations are identical when the thermal diffusivity α is equal to the kinematic viscosity v. The temperature can be nondimensionalized with respect to the temperature difference between the wall and the fluid

$$\theta \equiv \frac{T - T_s}{T_\infty - T_s}$$

(4-95)

and the velocities can be nondimensionalized by dividing V_∞, making the thermal boundary condition $\theta = 0$ at $y = 0$ analogous to $u/V_\infty = 0$ at $y = 0$. Temperature and velocity profiles are identical when the dimensionless *Prandtl number*,

$$Pr \equiv \frac{v}{\alpha} \tag{4-96}$$

is unity, which is approximately the case for most gases ($0.6 < Pr < 1.0$). The Prandtl number for liquids, however, varies widely, ranging from very large values for viscous oils to very small values (on the order of 0.01) for liquid metals which have high thermal conductivities.

4.4.5.2   Exact Energy Solution (Pohlhausen Solution)

The similarity between the momentum and energy equations led Pohlhausen to use a parameter and stream function which incorporates all unknowns into one variable,

$$\eta = y \sqrt{\frac{V_\infty}{vx}} \qquad\qquad \psi = \sqrt{vxV_\infty} \; f(\eta) \tag{4-97}$$

giving the ordinary linear differential equation

$$\frac{d^2\theta}{d\eta^2} + \frac{Pr}{2} f \frac{d\theta}{d\eta} = 0 \tag{4-98}$$

with boundary conditions $\theta(0) = 0$, $\theta(\infty) = 1$. The solution is

$$\theta(\eta) = \frac{\displaystyle\int_0^\eta \exp\left(-\frac{Pr}{2} \int_0^\beta f(\alpha)\, d\alpha\right) d\beta}{\displaystyle\int_0^\infty \exp\left(-\frac{Pr}{2} \int_0^\beta f(\alpha)\, d\alpha\right) d\beta} \tag{4-99}$$

where $f(\alpha)$ is known (in numerical form at least) from the Blasius solution. The temperature distribution Eq. (4-99) is plotted in Figure 4.9 for several Prandtl numbers; the curve for $Pr = 0.7$ is typical for air and several other gases.

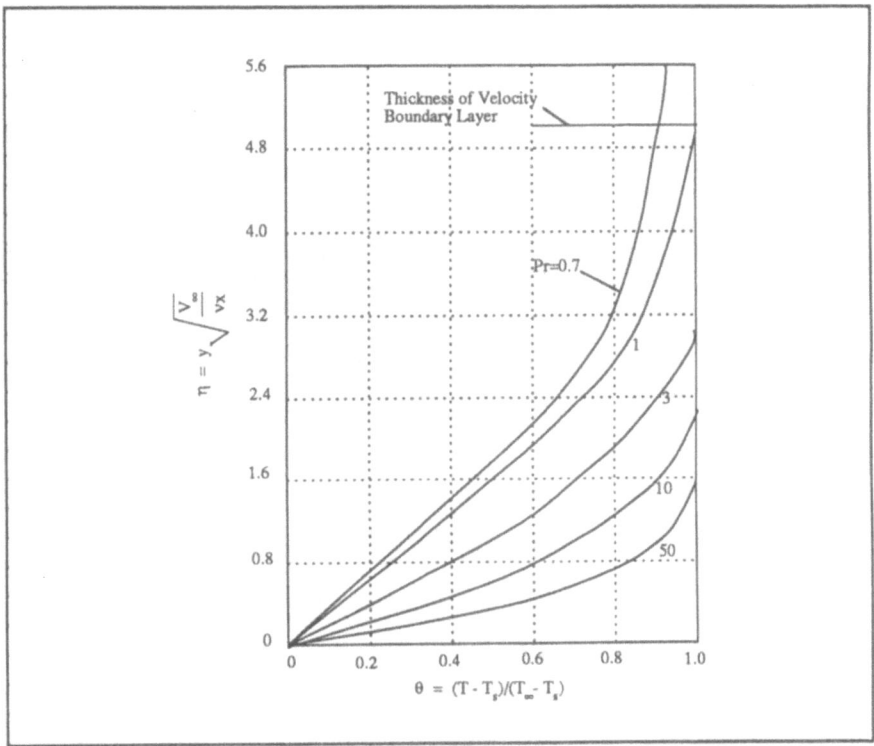

FIGURE 4.9   Temperature Distribution Inside the Boundary Layer for Various Prandtl Numbers

The slope of the temperature profile at the surface, y = 0, is well represented by

$$\left(\frac{\partial T}{\partial y}\right)_{y=0} = (T_\infty - T_s) \sqrt{\frac{V_\infty}{vx}} \left(\frac{d\theta}{d\eta}\right)_{\eta=0} \approx (T_\infty - T_s) \sqrt{\frac{V_\infty}{vx}} (0.332) \, Pr^{1/3}$$

$$(4\text{-}100)$$

for $0.6 < \mathbf{Pr} < 15$. Substitution of this expression into (4-98) yields

$$h = 0.332k \sqrt{\frac{V_\infty}{vx}} \, Pr^{1/3}$$

$$(4\text{-}101)$$

Multiplying through by x/k, we get the local dimensionless *Nusselt number*,

$$\mathbf{Nu_x} \equiv \frac{h_x x}{k} = (0.332) \, \mathbf{Re_x^{1/2}} \, \mathbf{Pr^{1/3}}$$

$$(4\text{-}102)$$

Taking averages over the interval $0 < x < L$, the average heat-transfer coefficient and Nusselt number are found to be

$$\overline{h} = 2(h_x)_{x=L} \tag{4-103}$$

$$\overline{Nu} \equiv \frac{\overline{h}L}{k} = (0.664)\, Re_L^{1/2}\, Pr^{1/3} \tag{4-104}$$

4.4.5.3   Approximate Energy Solution (Integral Method)

The von Kármán integral technique can be applied to both the fluid and thermal boundary layers over a flat plate (Figs. 4.10 and 4.11). The momentum and energy balance for the boundary layer are:

Force balance at the boundary layer
$$\nu\left(\frac{\partial u}{\partial y}\right)_{y=0} = \frac{\partial}{\partial x}\int_0^\delta u\,(u - V_\infty)\,dy \tag{4-105}$$

Energy balance at the boundary layer
$$\alpha\left(\frac{\partial T}{\partial y}\right)_{y=0} = \frac{\partial}{\partial x}\int_0^{\delta_t} u\,(T_\infty - T)\,dy \tag{4-106}$$

FIGURE 4.10   Momentum Balance at the Boundary Layer

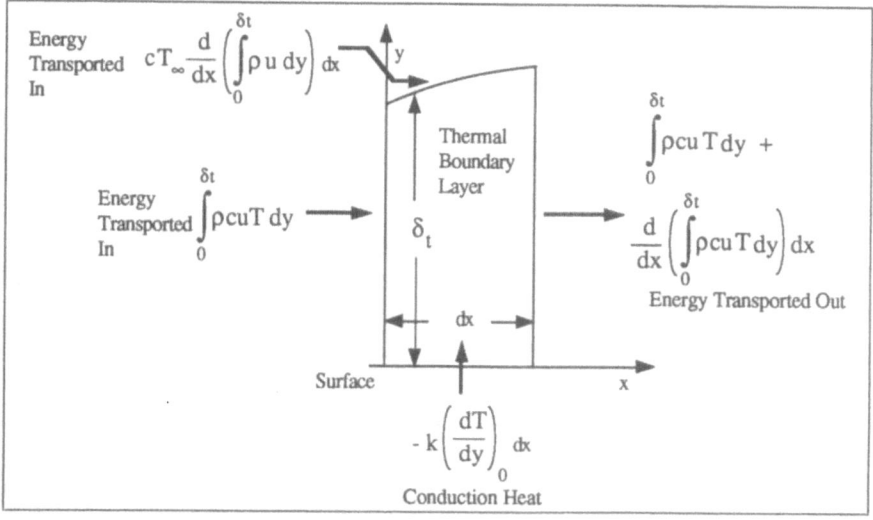

FIGURE 4.11   Energy Balance at the Boundary Layer

To solve Eq. (4-106) for δ_t we need to know, or to assume, a velocity profile and a temperature distribution within the boundary layer that respectively satisfy the boundary conditions For Eq. (4-99). If third-degree polynomials are used for u/V_∞ and θ, the resultant local Nusselt number for convective heat-transfer from a flat plate is

$$\mathbf{Nu}_x \equiv \frac{h_x x}{k} = (0.332)\, \mathbf{Re}_x^{1/2}\, \mathbf{Pr}^{1/3} \left[1 - \left(\frac{x_i}{x}\right)^{3/4}\right]^{-1/3}$$

(4-107)

Here, x_i is the length of an unheated leading section of the plate; when $x_i = 0$, Eq. (4-107) is identical to the Pohlhausen solution, Eq. (4-104).

4.4.6 Turbulent Flow

Turbulent flow is characterized by random motion of fluid particles, disrupting the fluid's movement in lamina as discussed in the preceding chapters. It is the most common type of motion, however, because only minimal disturbances are needed to initiate turbulent flow.

The velocity in turbulent flow consists of the average value $\overline{V}$ and a fluctuating part V':

$$V = \overline{V} + V'$$

(4-108)

Taking the time average over a long period of time, the following is obtained:

$$\overline{V} = \frac{\lim}{\Delta t \to \infty} \frac{1}{\Delta t} \int_{t_0}^{t+\Delta t} V \, dt$$

(4-109)

i.e. the fluctuations cancel out in the long term. Any fluid property, say viscosity μ, may be similarly time-averaged: $\mu = \overline{\mu} + \mu'$.

The time-average equation for velocity, Eq. (4-108), is valid for any quantity ϕ which has a time-average value $\overline{\phi}$ and a fluctuating component ϕ', i.e. $\phi = \overline{\phi} + \phi'$. Thus

$$\overline{\phi} = \frac{\lim}{\Delta t \to \infty} \frac{1}{\Delta t} \int_{t_0}^{t+\Delta t} \phi \, dt$$

(4-110)

where $\Delta t = t - t_0$. Properties of the time average are:

$$\overline{\phi_1 + \phi_2} = \overline{\phi_1} + \overline{\phi_2}$$

(4-111)

$$\overline{C\phi_1} = C\overline{\phi_1}$$

(4-112)

$$\overline{\overline{\phi_1}} = \overline{\phi_1}$$

(4-113)

$$\overline{\frac{\partial \phi_1}{\partial s}} = \frac{\partial \overline{\phi_1}}{\partial s}$$

(4-114)

$$\overline{\phi'_1} = 0$$

(4-115)

where C is independent of t and s is any spatial coordinate. In addition, it is almost always the case that $\overline{\phi'_1 \phi'_2} \neq 0$ if ϕ_1 and ϕ_2 are turbulent flow properties.

4.4.6.1   Equations of Motion

By use of the boundary layer concept, the general equations of motion, called the *Navier-Stokes equations* after their formulators, can be simplified to the point of being solved. The x-direction momentum equation for incompressible, laminar, boundary layer flow over a flat plate is a simplified form of the more general x-direction Navier-Stokes equation and can be extended to the case of turbulent flow, i.e.

laminar:
$$u\frac{\partial u}{\partial x} + v\frac{\partial u}{\partial y} = -\frac{1}{\rho}\frac{dp}{dx} + v\frac{\partial^2 u}{\partial y^2}$$
(4-116)

turbulent:
$$(\bar{u} + u')\frac{\partial}{\partial x}(\bar{u} + u') + (\bar{v} + v')\frac{\partial}{\partial y}(\bar{u} + u')$$

$$= -\frac{1}{\rho}\frac{dp}{dx}(\bar{p} + p') + v\frac{\partial^2}{\partial y^2}(\bar{u} + u')$$
(4-117)

Velocity and pressure in the laminar equation have been divided into their average and fluctuating components in the turbulent equation.

Along with the momentum equations, the two-dimensional incompressible continuity equation may be considered:

laminar:
$$\frac{\partial u}{\partial x} + \frac{\partial v}{\partial y} = 0$$
(4-118)

turbulent:
$$\frac{\partial}{\partial x}(\bar{u} + u') + \frac{\partial}{\partial y}(\bar{v} + v') = 0$$
(4-119)

Combining the turbulent equations (4-117) and (4-119) and taking the time average of the resultant equation, the following is obtained:

$$\bar{u}\frac{\partial \bar{u}}{\partial x} + \bar{v}\frac{\partial \bar{u}}{\partial y} = -\frac{1}{\rho}\frac{d\bar{p}}{dx} + v\frac{\partial^2 \bar{u}}{\partial y^2} - \left(\overline{\frac{\partial u'u'}{\partial x}} + \overline{\frac{\partial v'u'}{\partial y}}\right)$$
(4-120)

which is the x-direction equation of motion for a viscous incompressible fluid with negligible body forces. This equation can be expected to hold in the turbulent portion of the boundary layer shown in Figure 4.12 (a). The laminar portion at the leading edge is governed by the relations developed earlier in this chapter. This chapter deals with the turbulent regime, which is often idealized as shown in Figure 4.12 (b), in which the critical length x_c is taken as that distance required to produce a Reynolds number of 500,000 (although this may vary from 300,000 to 2,800,000 depending upon such factors as surface roughness and free-stream turbulence).

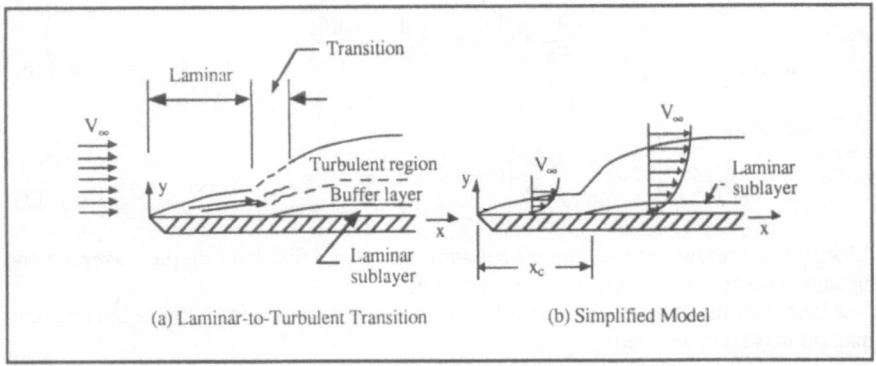

FIGURE 4.12   Turbulent Boundary Layer

Upon comparing Eq. (4-120) with the laminar equation (4-78), note that an additional term occurs in the equation for turbulent flow. The three terms on the right-hand side of Eq. (4-120) express the effects of pressure (normal stress), viscosity (shear stress), and turbulent fluctuations (*Reynolds*, or *apparent*, *stress*). The last term produces an apparent force

$$(f_x)_{apparent} = - \rho \left(\frac{\partial \overline{u'u'}}{\partial x} + \frac{\partial \overline{v'u'}}{\partial y} \right)$$

(4-121)

which will be interpreted in the next subsection.

4.4.6.2   Eddy Viscosity

Assume turbulent motion along the surface of Figure 4.13. At a typical plane parallel to the surface, A-A, a lump of fluid designated 1 is moved by turbulent fluctuations to a region of increased velocity. It is replaced by lump 2 moving downward to a region of lower velocity. Since momentum is the product of mass and velocity, a change in momentum results. This change can be evaluated quantitatively by considering the velocity components

$$u = \overline{u} + u' \qquad v = 0 + v'$$

(4-122)

where $\overline{v} = 0$ results from zero mean flow in the y-direction. The instantaneous mass flow per unit area across plane A-A is $\rho v'$. Multiplying by the x-velocity deviation u' gives the rate of momentum change per unit area $\rho v'u'$, whose average value, $\overline{\rho v'u'}$, is negative. Thus there is a shear stress $-(\rho) \overline{v'u'}$ over and above the laminar shear $\mu_m (\partial \overline{u}/\partial y)$, i.e.

$$\tau = \tau_{lam} + \tau_{eddy} = \left(\mu_m \frac{\partial \overline{u}}{\partial y} - \rho \overline{v'u'} \right)$$

(4-123)

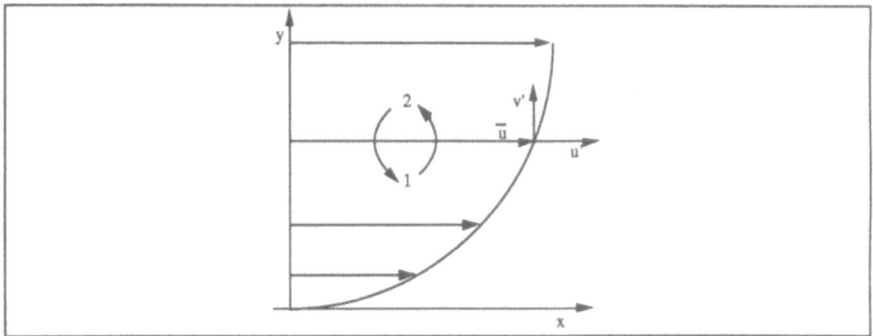

FIGURE 4.13   Fluid Mixing in the Turbulent Boundary Layer

This result is not readily usable, except through measurement of the fluctuating components, since the fluctuating velocities are not related to fluid properties. It is convenient to assume that the viscosity effect is increased because of the turbulence and define an *eddy viscosity* ε such that

$$\tau = (\mu_m + \rho\varepsilon) \frac{\partial \overline{u}}{\partial y}$$

(4-124)

or, since $v = \mu_m/\rho$,

$$\frac{\tau}{\rho} = (v + \varepsilon) \frac{\partial \overline{u}}{\partial y}$$

(4-125)

Similarly, there is an additional normal stress

$$\tau_x = \left(\mu_m \frac{\partial \overline{u}}{\partial x} + \rho \overline{u'u'} \right)$$

(4-126)

The time-averaged products of the fluctuating components, $\overline{v'u'}$ and $\overline{u'u'}$, which arose in Eq. (4-120), produce shear and normal stresses as shown in Eq. (4-123) and (4-126). It is then only necessary to measure the eddy viscosity, which is directional-dependent, in order to effect a solution to the equation of motion.

4.4.6.3 Eddy Diffusivity

An argument analogous to that of the preceding subsection can be made for the influence of turbulence on heat transfer. If a temperature fluctuation exists in the flow field of Figure 4.13, so that $T = \overline{T} + T'$, the heat transfer per unit area would be given by

$$q'' = q''_{lam} + q''_{turb} = -k\frac{\partial \overline{T}}{\partial y} + \rho c_p \overline{v'T'} = -(k + \rho c_p \varepsilon_H)\frac{\partial \overline{T}}{\partial y} \quad (4\text{-}127)$$

or, since $\alpha = k/\rho c_p$,

$$\frac{q''/c_p}{\rho} = -(\alpha + \varepsilon_H)\frac{\partial \overline{T}}{\partial y} \quad (4\text{-}128)$$

where the time-averaged product of the fluctuating velocity v' and fluctuating temperature T' has been replaced by an *eddy diffusivity for heat*, ε_H. Equations (4-125) and (4-128) are now of identical form, which suggests the analogy between momentum transfer and heat transfer in the following section.

4.4.7   Heat Transfer in Turbulent Flow

In the vicinity of a surface, the fluid is essentially stationary, requiring that the transfer of heat take place by conduction. Such is the case in the laminar sublayer, where the heat transfer and shear stress are in the ratio

$$\left(\frac{q''}{\tau}\right)_{lam} = \frac{-k\frac{\partial T}{\partial y}}{\mu_m \frac{\partial u}{\partial y}} = -\frac{k}{\mu_m}\left(\frac{dT}{du}\right)_{x\,=\,cons} \quad (4\text{-}129)$$

At some distance from the surface, where the random fluctuations transport momentum and heat, the turbulent shear stress and heat transfer (at fixed x) have the instantaneous values

$$\tau_{turb} = -\rho\, v'\Delta u \qquad q''_{turb} = \rho c_p v'\Delta T \quad (4\text{-}130)$$

giving, in the limit,

$$\left(\frac{q''}{\tau}\right)_{turb} = -c_p\left(\frac{dT}{du}\right)_{x\,=\,cons} \quad (4\text{-}131)$$

A comparison of Eq. (4-129) and (4-131) shows that when $k/\mu_m = c_p$, i.e. when

$$Pr \equiv \frac{\mu_m c_p}{k} = 1 \quad (4\text{-}132)$$

then the single equation

$$\frac{q''}{\tau} = -c_p \left(\frac{dT}{du}\right)_{x \,=\, cons}$$ (4-133)

is valid through the entire boundary layer. T and u can be assumed to change at proportional rates through the boundary layer. Then each side of Eq. (4-133) is a constant, i.e.

$$\frac{q''}{\tau} = \text{constant} = \frac{q_s''}{\tau_s} = -c_p \left(\frac{dT}{du}\right)_{x \,=\, cons}$$ (4-134)

where q_s'' and τ_s are the surface values. Integrating through the boundary layer and applying the known surface boundary condition (u = 0, T = T_s), the following is derived:

$$-\frac{1}{c_p}\frac{q_s''}{\tau_s}\bigg|\int_0^{''} du = \int_{T_s} dT$$ (4-135)

The condition **Pr** = 1 necessary for Eq. (4-135) is approximated by many real fluids. The upper limits of integration in Reynolds' analogy will depend upon the flow configuration.

4.4.7.1 Shear Stress and Friction Factor

The integral momentum equation, Eq. (4-105), is equally valid for turbulent flow since no flow assumption was required when choosing the control volume. Therefore, in the incompressible case,

$$\tau_s = \rho \frac{d}{dx}\left[\int_0^\delta \left(V_\infty - \bar{u}\right)\bar{u}\,dy\right]$$ (4-136)

Notice that the mean turbulent velocity $\bar{u}$ is used over the whole range of integration; the laminar sublayer and the buffer zone are neglected.

In the fully turbulent portion of the boundary layer the velocity increases approximately as the one-seventh power of distance from the wall, giving

$$\frac{\bar{u}}{V_\infty} = \left(\frac{y}{\delta}\right)^{1/7}$$ (4-137)

which is known as the *one-seventh-power law*. For the same regime, Blasius experimentally deduced that the shear stress is related to the boundary layer thickness by

$$\tau_s = (0.0225)\rho V_\infty^2 \left(\frac{\nu}{V_\infty \delta}\right)^{1/4}$$ (4-138)

for Reynolds numbers ranging from 5×10^5 to 10^7. Using Eq. (4-137) and (4-138) in Eq. (4-136), the boundary layer thickness and the local skin-friction coefficient are found to be

$$\frac{\delta}{x} = \frac{0.376}{(V_\infty x/\nu)^{1/5}} = \frac{0.376}{\mathrm{Re}_x^{1/5}}$$

(4-139)

$$c_f \equiv \frac{\tau_s}{\rho V_\infty^2/2} = \frac{0.0576}{\mathrm{Re}_x^{1/5}}$$

(4-140)

and the average skin-friction coefficient is

$$c_{f,avg} \equiv \frac{F_f}{(\rho V_\infty^2/2)L} = \frac{0.072}{\mathrm{Re}_x^{1/5}}$$

(4-141)

where F_f is the friction drag per unit width.

These equations make no allowance for the laminar boundary layer, $0 < x < x_c$, which precedes the turbulent portion. They are quite accurate, however, beyond the critical length x_c when the length is taken as if the turbulent boundary layer begins at the leading edge of the plate. Both laminar and turbulent drag can be accounted for by subtracting the turbulent drag for the critical length and adding the laminar drag for that portion, i.e.

$$C_f = \frac{0.072}{\mathrm{Re}_x^{1/5}} - \frac{0.072}{\mathrm{Re}_{x_c}^{1/5}}\frac{x_c}{L} + \frac{1.328}{\sqrt{\mathrm{Re}_{x_c}}} = \frac{x_c}{L}$$

(4-142)

For a critical Reynolds number of 5×10^5, Eq. (4-142) simplifies to

$$C_f = \frac{0.072}{\mathrm{Re}_L^{1/5}} - (0.00334)\frac{x_c}{L}$$

(4-143)

4.4.7.2   Reynold's Analogy

At the edge of the boundary layer $u = V_\infty$ and $T = T_\infty$, which permits the upper limits of integration to be added to Eq. (4-135). The integration yields

$$-\frac{1}{c_p}\frac{q_s''}{\tau_s} V_\infty = T_\infty - T_s \quad \text{or} \quad \frac{-q_s''}{T_\infty - T_s}\frac{1}{\rho c_p V_\infty} = \frac{1}{2}\frac{\tau_s}{\rho V_\infty^2/2}$$

(4-144)

substitution of Eq. (4-140) into Eq. (4-144) yields

$$\frac{h_x}{\rho c_p V_\infty} = \frac{c_f}{2}$$

$$(4\text{-}145)$$

The dimensionless group of terms on the left side of Eq. (4-145), called the *Stanton number*, **St**, is the Nusselt number divided by the product of the Reynolds and Prandtl numbers, i.e.

$$\frac{Nu_x}{Re_x Pr} \equiv St_x = \frac{c_f}{2}$$

$$(4\text{-}146)$$

which is Reynolds' analogy for the flat plate, relating the skin friction to the heat transfer. A. P. Colburn showed that the analogy may be modified to pertain to Prandtl numbers ranging from 0.6 to 50 by

$$j_H \equiv St_x \ Pr^{2/3} = \frac{c_f}{2}$$

$$(4\text{-}147)$$

where j_H is known as the *Colburn factor*, or simply the *j-factor*, for heat transfer. Taking average values over $0 < x < L$, we obtain

$$\frac{\overline{h}}{\rho c_p V_\infty} = \frac{c_{f,avg}}{2}$$

$$(4\text{-}148)$$

$$\overline{j}_H = \overline{St} \ Pr^{2/3} = \frac{C_f}{2}$$

$$(4\text{-}149)$$

The approximate expression for the average skin-friction coefficient, either Eq. (4-141), (4-142) or (4-143), may be substituted into the Colburn equation (4-149) to get the average convective heat-transfer coefficient. Using Eq. (4-143) gives (for a critical Reynolds number of 500,000)

$$\overline{Nu} \equiv \frac{\overline{h}L}{k} = Pr^{1/3}(0.036 Re_L^{0.8} - 836)$$

$$(4\text{-}150)$$

Thus far, the laminar sublayer and the buffer zone have been ignored. Von Kármán, including both of these, found the local Stanton number for flow over a plane surface to be

$$St_x \equiv \frac{Nu_x}{Re_x \ Pr} = \frac{c_f/2}{1 + 5\sqrt{c_f/2}\left[(Pr - 1) + \ln\left(\frac{5Pr + 1}{6}\right)\right]}$$

$$(4\text{-}151)$$

which results to Reynolds' analogy for $Pr = 1$. If Eq. (4-140) is introduced into Eq. (4-151), the result is

$$\mathbf{Nu}_x \equiv \frac{h_x x}{k} = \frac{(0.0288)\,\mathbf{Re}_x^{0.8}\,\mathbf{Pr}}{1 + (0.849)\,\mathbf{Re}_x^{-0.10}\left[(\mathbf{Pr}-1) + \ln\left(\frac{5\mathbf{Pr}+1}{6}\right)\right]} \quad (4\text{-}152)$$

In all the equations of this section the pertinent parameters should be evaluated at the film temperature, $T_f = (T_s + T_\infty)/2$.

4.4.8   Pipe Flow

Since in laser machining, the aspect ratio of the kerf, hole or groove is large, an analogy between fluid flow in the kerf and pipe flow can often be made. If the flat plate studied in the earlier sections is rolled into a pipe, then the concepts from boundary layer and turbulence flow for flat plates can be applied to flow in pipes. One difference between the two cases is that while the boundary layer continues to increase in thickness as a fluid passes along a flat plate, the boundary layer thickness in a pipe is physically limited to the radius of the pipe.

Figure 4.14 shows the successive stages of development of the boundary layer of an incompressible viscous fluid in the entrance region of a circular tube. At the tube entrance uniform flow at the free-stream velocity, exists. As the fluid moves down the tube, shear between the fluid and the wall, and between adjacent fluid particles, retards the motion, causing the boundary layer to grow until it is *fully developed* at Station 3. From this point on, the velocity profile remains unchanged.

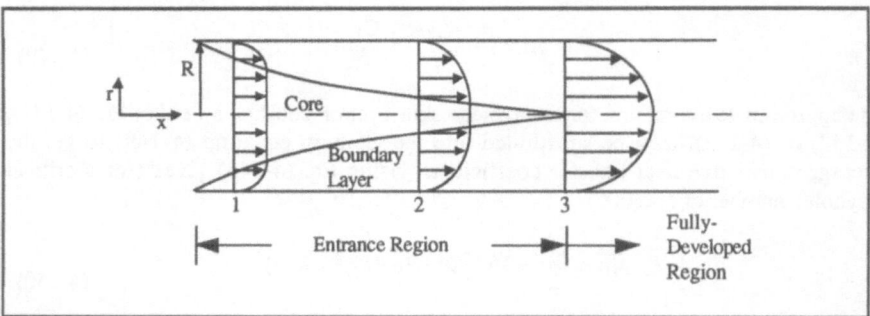

FIGURE 4.14   Boundary Layer Development in a Circular Tube

4.4.8.1   Entrance Region

The length required for the velocity profile to become invariant with axial position is known as the *entry length*, x_e. For the case of laminar flow, it may be approximated by the simple *Langhaar equation*

$$x_e \approx (0.05)\,\mathbf{Re}_D D \quad (4\text{-}153)$$

In most cases the entry length of a pipe is negligible when compared with its total length. Most engineering calculations are, therefore, made assuming fully developed flow throughout.

4.4.8.2   Fully-Developed Region

The fully-developed velocity profile for pipe flow can be derived from the Navier-Stokes equation (4-77). For steady, fully developed, laminar flow in a tube, the velocity profile is parabolic:

$$u = -\frac{1}{4\mu_f}\frac{dp}{dx}\left(R^2 - r^2\right)$$
(4-154)

where the minus sign is required because the pressure decreases in the flow direction. The maximum velocity, which occurs when $r = 0$, is

$$V_0 = -\frac{1}{4\mu_f}\frac{dp}{dx}R^2$$
(4-155)

The average velocity, V, may be obtained by equating the volumetric flow to the integrated paraboloidal flow, i.e.

$$V\pi R^2 = \int_0^R u\,(2\pi r)\,dr$$
(4-156)

giving

$$V = \frac{V_0}{2} = -\frac{1}{8\mu_f}\frac{dp}{dx}R^2$$
(4-157)

The pressure gradient can be expressed in terms of a *friction factor*, f, defined by

$$-\frac{dp}{dx} = \frac{f}{D}\frac{\rho V^2}{2}$$
(4-158)

where D is the tube diameter. Integration of this expression results in the *Darcy-Weisbach equation*,

$$\frac{\Delta p}{L} = \frac{f}{D}\frac{\rho V^2}{2}$$
(4-159)

where $\Delta p = p_1 - p_2$ and $L = x_2 - x_1$. The head loss between Stations 1 and 2 is $h_L = (p_1 - p_2)/\rho g$; thus,

$$h_L = f \frac{L}{D} \frac{V^2}{2g}$$
(4-160)

Eq. (4-157) and (4-158) give a simple relationship between the friction factor and the Reynolds number.

$$f = \frac{64}{Re_D}$$
(4-161)

which is valid for laminar tube flow, $Re < 2000$.

In terms of volumetric flow rate, $Q = AV$, the head loss is given by

$$h_L = \frac{\Delta p}{\rho g} = 128 \frac{QL\mu_f}{\pi D^4 \rho g}$$
(4-162)

The volumetric flow rate may be conveniently expressed in terms of the pressure drop; this is known as the *Hagen-Poiseuille equation*.

$$Q = \frac{\pi}{8\mu_f} \frac{R^4}{L} (p_1 - p_2)$$
(4-163)

4.4.8.3   Noncircular Ducts

The Darcy-Weisbach equation, Eq. (4-159), is valid for noncircular ducts when the geometric diameter is replaced by the *hydraulic diameter* D_h defined by

$$D_h \equiv \frac{4A}{P}$$
(4-164)

where A is the cross-sectional area and P is the wetted duct perimeter. The friction factor can be evaluated for the specific duct configuration.

4.4.9   Heat Transfer in Pipe Flows

4.4.9.1   Heat Transfer in Laminar Pipe Flow

A large class of heat transfer problems of engineering importance involves the flow of fluids through pipes. The thermal boundary layer, which develops similarly to the hydrodynamic boundary layer shown in Figure 4.14, is significant in the heat transfer process. Although heat transfer in laminar flow is not very common because the rate is lower than that encountered in turbulent flow, it is sometimes desirable due to the lower pumping power required in the laminar case.

In purely laminar flow, the heat transfer mechanism is conduction, resulting in large heat-transfer coefficients for fluids with high thermal conductivities, such as liquid metals. Figure 4.15 shows the variation of the local heat-transfer coefficient (Nusselt number) with axial distance along a tube in developing laminar flow for $Pr = 0.7$. Three cases of boundary conditions are shown. A unique feature of these curves is the asymptotic values of the Nusselt number as the flow becomes fully developed.

constant heat flux:
$$Nu_{D\infty} \equiv \frac{h_\infty D}{k} = 4.364 \qquad (4\text{-}165)$$

constant wall temperature:
$$Nu_{D\infty} \equiv \frac{h_\infty D}{k} = 3.656 \qquad (4\text{-}166)$$

While the asymptotic Nusselt numbers can be applied to long pipes, some engineering applications involve tubes which are too short for the flow to fully develop. In these cases, the heat transfer must be analyzed over the length in which the flow develops, called the *thermal entry length* $x_{e,t}$. The thermal entry length for laminar flows can be related to the Reynolds number and the Prandtl number through the equation

$$x_{e,t} \approx 0.05 \, Re_D \, Pr_D \qquad (4\text{-}167)$$

An analytical formula that supplements the results of Figure 4.15 was developed by H. Hausen:

uniform heat flux:
$$Nu_D = Nu_{D\infty} + \frac{K_1 \left[(D/x) \, Re_D Pr \right]}{1 + K_2 \left[(D/x) \, Re_D Pr \right]^n} \qquad (4\text{-}168)$$

constant wall temperature:
$$\overline{Nu}_D = Nu_{D\infty} + \frac{K_1 \left[(D/x) \, Re_D Pr \right]}{1 + K_2 \left[(D/x) \, Re_D Pr \right]^n} \qquad (4\text{-}169)$$

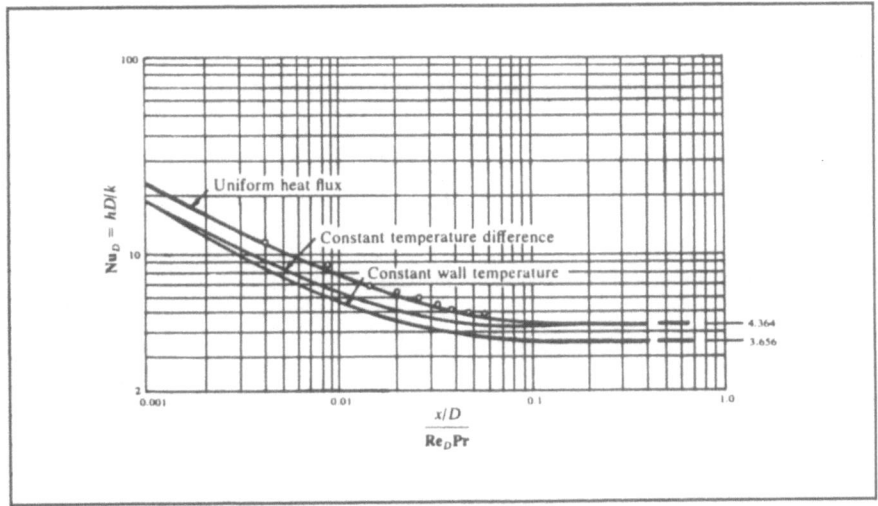

FIGURE 4.15   Variation of Heat Transfer Coefficient With Distance in a Circular Tube [11]

Boundary Condition	Inlet Velocity	Prandtl Number	Nusselt Number	K_1	K_2	n
Const. q/A	Parabolic	any	4.36	0.023	0.0012	1.0
Const. q/A	Developing	0.7	4.36	0.036	0.0011	1.0
Const. T	Parabolic	any	3.66	0.0668	0.04	0.67
Const. T	Developing	0.7	3.66	0.104	0.016	0.8

TABLE 4.1   Parametric Values in Heat Transfer Solutions for Laminar Pipe Flows

Nu_D corresponds to the local heat-transfer coefficient, while $\overline{Nu}_D$ corresponds to the heat transfer coefficient averaged over the tube length. Table 4.1 lists the parametric values to be used in Eq. (4-168) and (4-169) corresponding to four common sets of inlet conditions. For oils, or other fluids in which the viscosity varies considerably with temperature, the K_1-term in Eq. (4-168) can be modified empirically by multiplying with $(\mu/\mu_s)^{0.14}$, where μ is the fluid viscosity at ambient temperature and μ_s is the fluid viscosity at the wall temperature.

Except for μ_s, which is evaluated at the wall temperature, all fluid properties are evaluated at the *bulk temperature* T_b, which is the temperature that would be obtained if the fluid at a given cross section were directed into an insulated chamber and allowed to reach equilibrium. This definition makes the basic convection equation

$$q = hA(T_s - T_b) = mc_p(T_{b,o} - T_{b,i}) \qquad (4\text{-}170)$$

where $T_{b,i}$ and $T_{b,o}$ are the bulk temperatures at the inlet and outlet.

Mathematically, the bulk temperature is the enthalpy-average temperature of the fluid:

$$T_b \int_0^R \rho c_p u \, 2\pi r \, dr = \int_0^R \rho c_p u T \, 2\pi r \, dr \qquad (4\text{-}171)$$

which, for an incompressible fluid having constant specific heat, reduces to

$$T_b = \frac{\int_0^R uTr \, dr}{\int_0^R ur \, dr} \qquad (4\text{-}172)$$

In engineering practice, a simple approximate value,

$$T_b = \frac{T_{inlet} + T_{outlet}}{2} \qquad (4\text{-}173)$$

is used in the calculation of average heat-transfer coefficients.

4.4.9.2 Heat Transfer in Turbulent Pipe Flow

From a series of experiments, Nikuradse concluded that the velocity distribution for turbulent flow in smooth pipes is of the power-law form, i.e.

$$\frac{u}{V_{max}} = \left(\frac{y}{R}\right)^{1/n}$$

(4-174)

where u is the local time-average velocity, V_{max} is the time-average velocity at the centerline, R is the pipe radius and $y = R - r$ is the distance from the pipe wall. (For simplicity we will omit the overbar denoting the time average). Table 4.2 gives the values of n for several Reynolds numbers. Note that for **Re** = 1.1×10^5 the exponent is 1/7, making the profile of identical form to that of Eq. (4-137) for flow over a flat plate.

Re	n
4.0×10^3	6.0
2.3×10^4	6.6
1.1×10^5	7.0
1.1×10^6	8.8
2.0×10^6	10.0
3.2×10^6	10.0

TABLE 4.2   n Values for the Solution of Eq. (4-174)

The Darcy-Weisbach equation, Eq. (4-159), is equally valid for turbulent flow, but the friction factor f must be determined experimentally, rather than analytically as for laminar flow. In this equation V is the average velocity over the cross section, which for the power-law profile is expressed by

$$\frac{V}{V_{max}} = \frac{2n^2}{(2n + 1)(n + 1)}$$

(4-175)

For $10,000 < \mathbf{Re_D} < 100,000$, the friction factor f is well represented by

$$f = (0.184)\, \mathbf{Re_D}^{-0.2}$$

(4-176)

beyond the dimensionless distance given by H. Latzko:

$$\left(\frac{L}{D}\right)_c = (0.623)\, \mathbf{Re_D}^{1/4}$$

(4-177)

This is the distance required in turbulent flow for the friction factor to become constant, which is much less than the 40 to 50 diameters required for the turbulent velocity profile to develop.

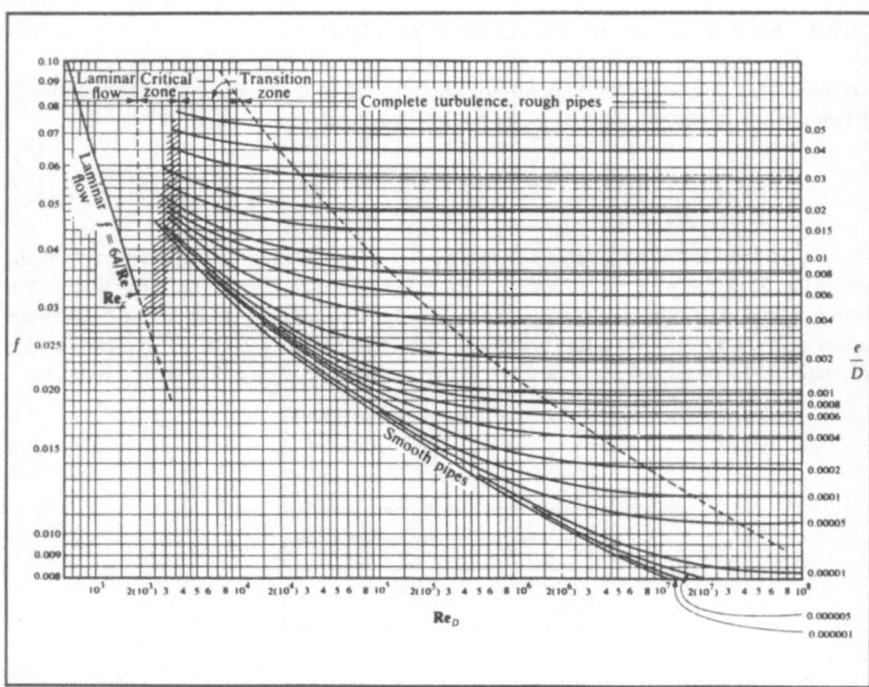

FIGURE 4.16   Moody's Diagram for Heat Transfer in Rough Pipes [11]

In rough pipes, where surface imperfections extend beyond the laminar sublayer, the friction factor f depends upon both Re_D and the *roughness height* e. Figure 4.16, commonly referred to as *Moody's diagram*, is a plot of friction factor versus Reynolds number, with the relative roughness e/D as parameter. Included are the results for hydraulically smooth pipes, discussed earlier in this section, as well as the straight-line, laminar flow relation (4-161). In the region of complete turbulence (high Reynolds number and/or large e/D) the friction factor depends predominantly upon the relative roughness, as shown by the flatness of the curves.

Applying the flat plate equation (4-135) for the case of turbulent flow in a tube (Fig. 4.17), the upper limits of integration are u = V, T = T_b. Here, V is the mean velocity and T_b is the bulk, or mean, temperature of the fluid:

$$T_b = \frac{T_i + T_o}{2}$$

(4-178)

where T_i and T_o are the average fluid temperatures at the inlet and outlet. Integrating (4-135), we get, analogous to (4-145),

$$\frac{h}{\rho c_p V} = \frac{\tau_s}{\rho V^2}$$

(4-179)

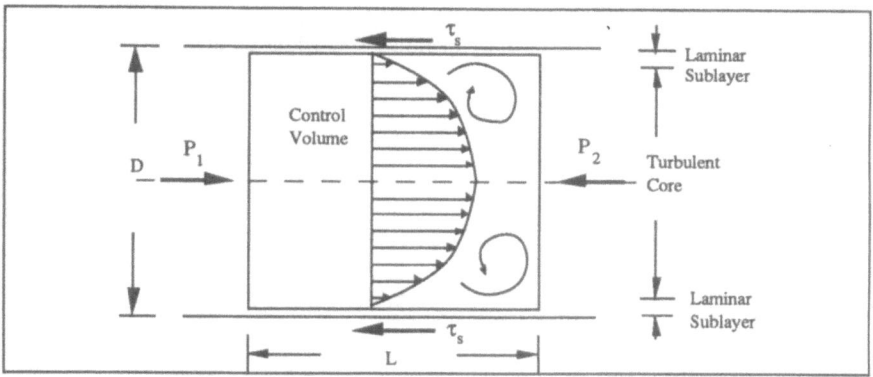

FIGURE 4.17   Force Balance for Turbulent Pipe Flow

A simple force balance on a cylindrical control volume of length L and diameter D gives

$$\tau_s = \frac{(p_1 - p_2)D}{4L} \qquad (4\text{-}180)$$

which can be combined with (4-159) to give

$$\tau_s = f\frac{\rho V^2}{8} \qquad (4\text{-}181)$$

Substituting Eq. (4-181) into (4-179), a relationship between the friction factor and the Stanton number can be derived.

$$St \equiv \frac{h}{\rho c_p V} = \frac{f}{8} \qquad (4\text{-}182)$$

Equations (4-182) is restricted to $Pr = 1$. The Colburn modification is applicable for the case of internal tube flow, giving

$$j_H = St \ Pr^{2/3} = \frac{f}{8} \qquad (4\text{-}183)$$

for fluids with Prandtl numbers from 0.5 to 100.

Substituting the expression (4-176) for f into (4-183) yields a working relation for Nusselt number:

$$Nu_D \equiv \frac{hD}{k} = (0.023) \ Re_D^{0.8} Pr^{1/3} \qquad (4\text{-}184)$$

valid for $10,000 < \text{Re}_D < 100,000$, $0.5 < \text{Pr} < 100$ and $L/D > 60$. The specific heat for Eq. (4-184) is evaluated at the bulk temperature T_b, while other fluid properties are evaluated at the average film temperature T_f, given by

$$T_f = \frac{T_s + T_b}{2} \tag{4-185}$$

A relation in which the fluid properties are evaluated at the bulk temperature T_b, making it much easier to use than (4-184), is the widely used *Dittus-Boelter equation*:

$$\text{Nu}_D \equiv \frac{hD}{k} = (0.023)\, \text{Re}_D^{0.8}\, \text{Pr}^n \tag{4-186}$$

where

$$n = \begin{cases} 0.4 \text{ for heating the fluid} \\ 0.3 \text{ for cooling the fluid} \end{cases} \tag{4-187}$$

This equation is valid for $10,000 < \text{Re}_D < 120,000$, $0.7 < \text{Pr} < 120$, and $L/D > 60$. Use of this equation should be limited to cases where the pipe surface temperature and the bulk fluid temperature differ by no more than $10°$ F for liquids and $100°$ F for gases.

For higher Prandtl numbers, $0.7 < \text{Pr} < 16,700$, and bigger temperature differences, the *Sieder-Tate equation*, which accounts for large changes in viscosity, is recommended.

$$\text{Nu}_D \equiv \frac{hD}{k} = (0.023)\, \text{Re}_D^{0.8}\, \text{Pr}^{1/3} \left(\frac{\mu_b}{\mu_s}\right)^{0.14} \tag{4-188}$$

Valid for $\text{Re}_D > 10,000$ and $L/D > 60$, this equation may be used for both heating and cooling. Except for μ_s, which is evaluated at the surface temperature, all properties are evaluated at the bulk temperature.

The equations of this section are valid for noncircular ducts when the duct diameter D is replaced by the hydraulic diameter D_h given by Eq. (4-164).

4.5 Radiation

Radiation is a term used for processes which involve energy transfer by electromagnetic wave phenomena. The radiative mode of heat transfer differs in two important respects from the conductive and convective modes: (1) no medium is required and (2) the energy transfer is proportional to the fourth or fifth power of the temperatures of the bodies involved.

Thermal radiation is defined as the portion of the electromagnetic spectrum between the wavelengths 1×10^{-7} m and 1×10^{-4} m (Fig. 4.18) or in the range between 0.1 and 100 μm. The value of any radiation quantity at a given wavelength is called a *monochromatic* value.

4.5.1 Radiative Surface Properties

Whenever radiant energy is incident upon any surface, part may be absorbed, part may be reflected, and part may be transmitted through the body. Defining

$\alpha \equiv$ fraction of incident radiation absorbed $\equiv$ absorptivity

$\rho \equiv$ fraction of incident radiation reflected $\equiv$ reflectivity

$\tau \equiv$ fraction of incident radiation transmitted $\equiv$ transmissivity

the sum of these three quantities must equal 1.

$$\alpha + \rho + \tau = 1 \qquad (4\text{-}189)$$

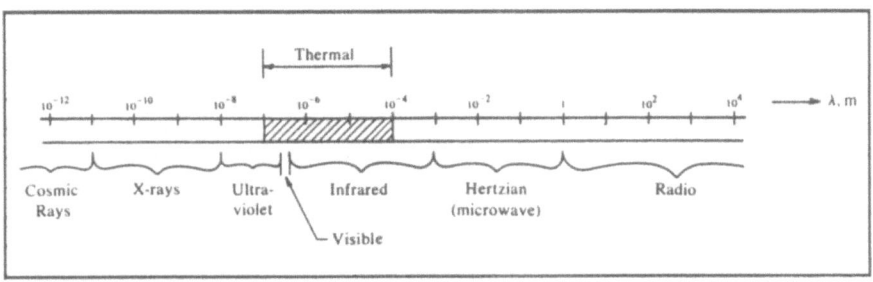

FIGURE 4.18   Spectrum of Electromagnetic Radiation [11]

Most solids, other than those which are visibly transparent or translucent, do not transmit radiation, and Eq. (4-189) reduces to

$$\alpha + \rho = 1 \qquad (4\text{-}190)$$

Frequently, (4-190) is applied to liquids, although the transmissivity of a liquid is strongly dependent upon thickness. The absorptivity value depends on the wavelength of the radiation, the surface morphology, etc.

Gases generally reflect very little radiant thermal energy, and (4-189) simplifies to

$$a + \tau = 1 \qquad (4\text{-}191)$$

4.5.2   Emissive Power and Radiosity

The *total emissive power*, denoted by E, is the total (over all wavelengths and all directions) emitted radiant thermal energy leaving a surface per unit time and unit area of the emitting surface. The emissive power does not include any energy reflected from the surface (and originating elsewhere). The total emissive power of a surface is dependent upon the material or substance, the surface condition and surface roughness, and the temperature.

Radiosity, J, denotes the total radiant thermal energy leaving a surface per unit time and unit area of the surface. Thus the radiosity is the sum of the emitted and the reflected radiant energy fluxes from a surface. Like total emissive power, total radiosity represents an integration over the spectral and directional distribution.

4.5.3   Specular and Diffuse Surfaces

Reflection of radiation from a surface can be either specular (where the radiation is reflected at an angle equal to the angle of incidence, such as for a mirror surface) or diffuse (where the incoming directional radiation is reflected in all directions). The degree of diffuse reflection depends on the surface roughness. This characteristic can be described with the help of two ideal models. For the case of a specular surface (Fig. 4.19a), the angle of incidence θ_i is equal to the angle made by the reflected ray, θ_r. Both θ_i and θ_r are measured from a line normal to the plane of the surface. For a diffuse surface (Fig. 4.19b), the incident ray is reflected hemispherically in all directions.

If the surface roughness for a real surface is considerably smaller than the wavelength of incident irradiation, the surface behaves as a specular reflector; if the roughness dimension is large with respect to wavelength, the surface reflects diffusely.

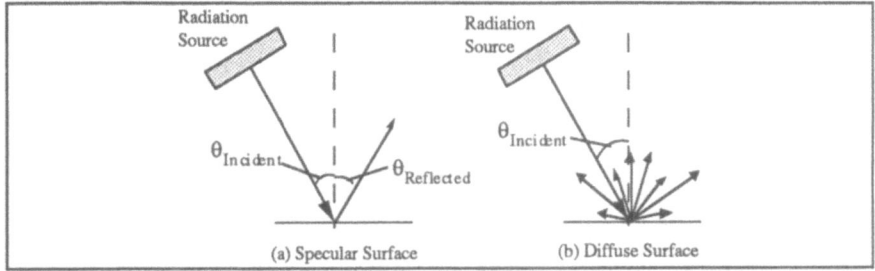

FIGURE 4.19   (a) Specular and (b) Diffuse Surfaces

4.5.4   Intensity of Radiation

The radiation intensity, I, is defined here as the radiant energy per unit time per unit solid angle per unit area of the emitter projected normal to the line of view of the receiver from the radiating element. For the geometry shown in Figure 4.20, the energy emitted from element dA_1 and intercepted by element dA_2 is

$$dq_{1\rightarrow 2} = I(\cos \phi \, dA_1)d\omega \qquad (4\text{-}192)$$

where

$$d\omega \equiv \frac{dA_2}{r^2} = \sin \phi \, d\theta \, d\phi \qquad (4\text{-}193)$$

is the solid angle subtended by dA_2, and $\cos \phi \, dA_1$ is the area of the emitting surface projected normal to the line of view to the receiving surface.

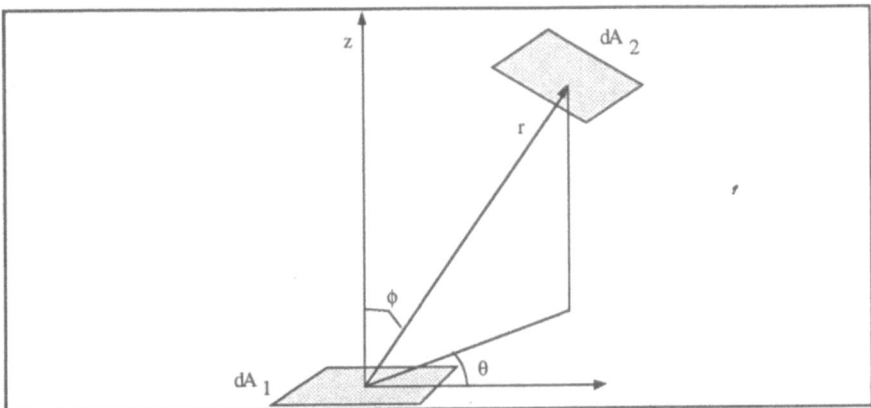

FIGURE 4.20   Energy Transfer Through Radiation Between Two Surfaces

Substituting (4-194) into (4-193) and integrating over the hemispherical surface results in

$$\frac{q_{1\rightarrow2}}{dA_1} = \int_0^{2\pi} \int_0^{\pi/2} I \cos\phi \sin\phi \, d\phi \, d\theta \tag{4-194}$$

which is the general relationship between the total emissive power of a body (in this case, the element dA_1) and the intensity of radiation.

If the emitting surface is diffuse, I = constant, and Eq. (4-194) integrates to

$$\frac{q_{1\rightarrow2}}{dA_1} = E = \pi I \tag{4-195}$$

4.5.5 Black Body Radiation

The ideal surface in the study of radiative heat transfer is the *black body*, which is defined by $\alpha_b = 1$. Thus the black body absorbs all incident thermal radiation, regardless of spectral or directional characteristics. Such a body can be approximated by a small hole leading into a cavity, called a Hohlraum (Fig. 4.21).

The total (hemispherical) emissive power of a black body is given by the *Stefan-Boltzmann equation*:

$$E_b = \sigma T^4 \tag{4-196}$$

where σ, the *Stefan-Boltzmann constant*, is 0.1714 x 10^{-8} Btu/hr-ft^2-°R^4 or 5.6697 x 10^{-8} W/m^2-K^4. Emission from a black body is independent of direction, so that Eq. (4-195) gives

$$E_b = \sigma T^4 = \pi I_b \tag{4-197}$$

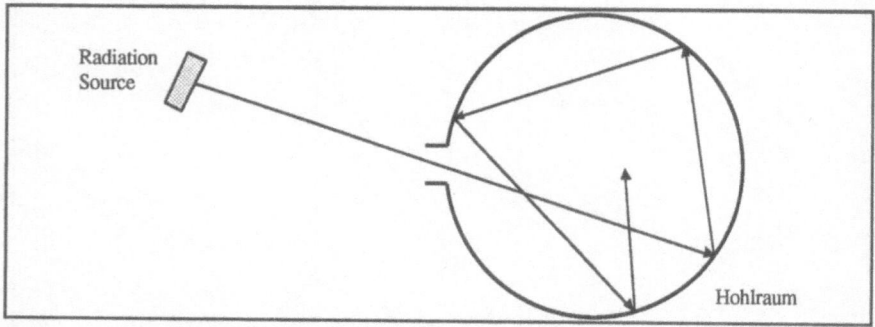

FIGURE 4.21   Configuration for a Hohlraum Black Body

In general, any surface emits different amounts of energy at different wavelengths. The total emissive power can be expressed as

$$E = \int_0^\infty E_\lambda \, d\lambda$$

(4-198)

where E_λ is the monochromatic emissive power at wavelength λ. For a black body,

$$E_b = \int_0^\infty E_{b\lambda} \, d\lambda = \sigma T^4$$

(4-199)

The first accurate expression for $E_{b\lambda}$ was determined by Max Planck; it is

$$E_{b\lambda} = \frac{C_1 \lambda^{-5}}{\exp\left(C_2/\lambda T\right) - 1}$$

(4-200)

in which

$$C_1 = 1.187 \times 10^8 \, \frac{\text{Btu-}\mu\text{m}^4}{\text{hr-ft}^2} = 3.742 \times 10^8 \, \frac{\text{W-}\mu\text{m}^4}{\text{m}^2}$$

(4-201)

$$C_2 = 2.5896 \times 10^4 \, \mu\text{m-}^\circ\text{R} = 1.4387 \times 10^4 \, \mu\text{m-K}$$

(4-202)

Plots of $E_{b\lambda}$ versus λ for several different temperatures are given in Figure 4.22. The shift in location of the maximum value of the monochromatic emissive power to shorter wavelengths with increasing temperature is evident. This wavelength shift is described by Wien's displacement law,

$$\lambda_{max} T = 5215.6 \, \mu\text{m-}^\circ\text{R} = 2897.6 \, \mu\text{m-K}$$

(4-203)

which plots as the dashed curve through the peak values of emissive power in Figure 4.22.

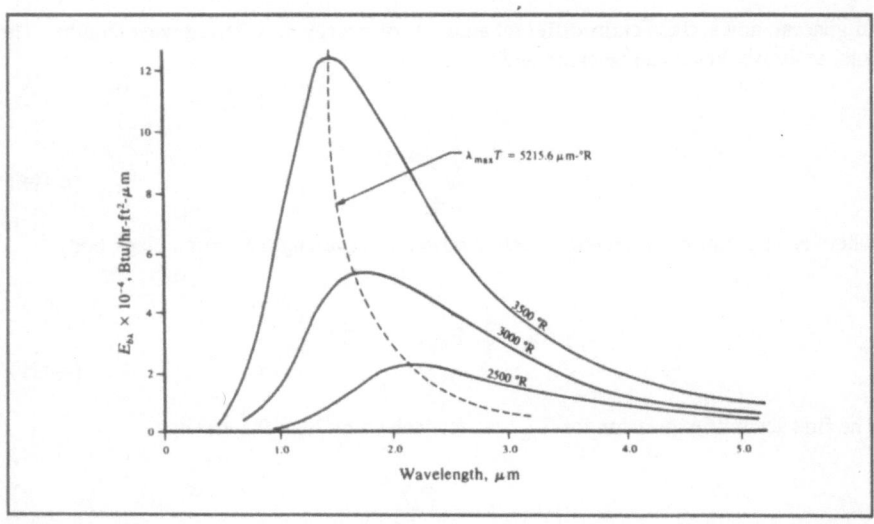

FIGURE 4.22   Emissive Power vs. Wavelength for Different Temperatures [11]

It is frequently necessary to determine the amount of energy radiated by a black body over a specified portion of the thermal radiation waveband. The energy emitted within the range 0 to λ at a specified temperature T can be expressed as

$$E_{b(0-\lambda T)} \equiv \int_0^{\lambda T} \frac{1}{T} E_{b\lambda} \, d(\lambda T)$$

(4-204)

The fraction of the total energy within this range is then

$$\frac{E_{b(0-\lambda T)}}{E_b} = \frac{E_{b(0-\lambda T)}}{\sigma T^4} = \int_0^{\lambda T} \frac{E_{b\lambda}}{\sigma T^5} \, d(\lambda T)$$

(4-205)

4.5.6 Grey Body Radiation

A real surface has a total emissive power E less than that of a black body. The ratio of the total emissive power of a body to that of a black body at the same temperature is the *total emissivity* (or *total hemispherical emissivity*), ε:

$$\varepsilon \equiv \frac{E}{E_b}$$

(4-206)

Some numerical values of total emissivity are presented in Table 4.3.

The *monochromatic (hemispherical) emissivity*, ε_λ, will be useful in dealing with real surfaces which exhibit spectrally selective emittance values. This is

$$\varepsilon_\lambda \equiv \frac{E_\lambda}{E_{b\lambda}}$$

(4-207)

where E_λ is the emissive power of the real surface at wavelength λ, and $E_{b\lambda}$ is that of a black body, both being at the same temperature.

The monochromatic emissive power of a real surface is not always directly proportional to that of a black surface; to simplify the computation process, however, a useful idealization of a real surface is that of a gray body, defined by

$$(\varepsilon_\lambda)_{gray} \equiv \text{constant}$$

(4-208)

The computational advantages of this are apparent from a consideration of the expression for the total emissive power of a body:

$$E = \int_0^\infty E_\lambda \, d\lambda = \int_0^\infty \varepsilon_\lambda E_{b\lambda} d\lambda$$

(4-209)

which for constant ε_λ simplified to

$$E = \varepsilon \int_0^\infty E_{b\lambda} \, d\lambda$$

(4-210)

or

$$E = \varepsilon \sigma T^4$$

(4-211)

In addition to the previously discussed variables which influence surface properties, the emissivity of a smooth surface depends strongly upon the polar angle ϕ between the direction of the incoming radiation and a normal to the surface. In general, nonconductors emit more strongly in the direction normal to the surface (or at small polar angles), whereas conductors emit more strongly at large polar angles.

Surface	T, °F	Emissivity ϵ
Metals and their oxides		
Aluminum:		
Highly polished plate, 98.3% pure	440–1070	0.039–0.057
Commercial sheet	212	0.09
Heavily oxidized	299–940	0.20–0.31
Brass:		
Highly polished:		
73.2% Cu, 26.7% Zn	476–674	0.028–0.031
62.4% Cu, 36.8% Zn, 0.4% Pb, 0.3% Al	494–710	0.033–0.037
82.9% Cu, 17.0% Zn	530	0.030
Hard-rolled, polished, but direction of polishing visible	70	0.038
Dull plate	120–660	0.22
Copper:		
Polished	242	0.023
	212	0.052
Plate, heated long time, covered with thick oxide layer	77	0.78
Gold, pure, highly polished	440–1160	0.018–0.035
Iron and steel (not including stainless):		
Steel, polished	212	0.066
Iron, polished	800–1880	0.14–0.38
Cast iron, newly turned	72	0.44
Cast iron, turned and heated	1620–1810	0.60–0.70
Mild steel	450–1950	0.20–0.32
Oxidized surfaces:		
Iron plate, pickled, then rusted red	68	0.61
Iron, dark-gray surface	212	0.31
Rough ingot iron	1700–2040	0.87–0.95
Sheet steel with strong, rough oxide layer	75	0.80
Lead:		
Unoxidized, 99.96% pure	260–440	0.057–0.075
Gray oxidized	75	0.28
Oxidized at 300 °F	390	0.63
Magnesium, magnesium oxide	530–1520	0.55–0.20
Molybdenum:		
Filament	1340–4700	0.096–0.202
Massive, polished	212	0.071
Monel metal, oxidized at 1110 °F	390–1110	0.41–0.46
Nickel:		
Polished	212	0.072
Nickel oxide	1200–2290	0.59–0.86
Nickel alloys:		
Copper nickel, polished	212	0.059
Nichrome wire, bright	120–1830	0.65–0.79
Nichrome wire, oxidized	120–930	0.95–0.98

*Abstracted by permission from H.C. Hottel in W.H. McAdams (ed.), *Heat Transmission*, 3d ed., pp. 472–478.   Copyright 1954, McGraw-Hill Book Company.

TABLE 4.3   Total Emissivity Values for Selected Materials

4.5.7 Radiative Exchange Between Two Bodies

The total radiant energy exiting a surface in all directions at all wavelengths is defined as the total emissive power. The emissive power is related to the fourth power of the absolute temperature. For the situation where two surfaces are emitting energy to each other, each surface will radiate at a magnitude equal to the total emissive power corresponding to the surface temperature (Fig. 4.23). The difference between the two total emissive powers for the two surfaces is the net radiative heat flux between these surfaces. The quantitative formulation of the net radiative heat flux can be given by Eq. (4-197). A black body emits energy at a rate proportional to the fourth power of the absolute temperature. When two black bodies exchange heat by radiation, the net heat exchange is

$$q_{rad} = \sigma(T_s^4 - T_o^4) \qquad (4\text{-}212)$$

where q_{rad} is radiative heat flux, and σ is the Stefan-Boltzmann constant. For two gray bodies, the net radiative heat flux is

$$q_{rad} = \varepsilon\sigma(T_s^4 - T_o^4) \qquad (4\text{-}213)$$

where ε is the emissivity of the material. Eqs. (4-212) and (4-213) both assume that the two surfaces exchange energy only with each other. In more complicated cases where a surface exchanges energy to more than one other surface, form factors must be calculated to determine the fraction of emitted energy which reaches each surface. Form factors depend on the size and orientation of each surface; details for form factor calculation are shown in [5].

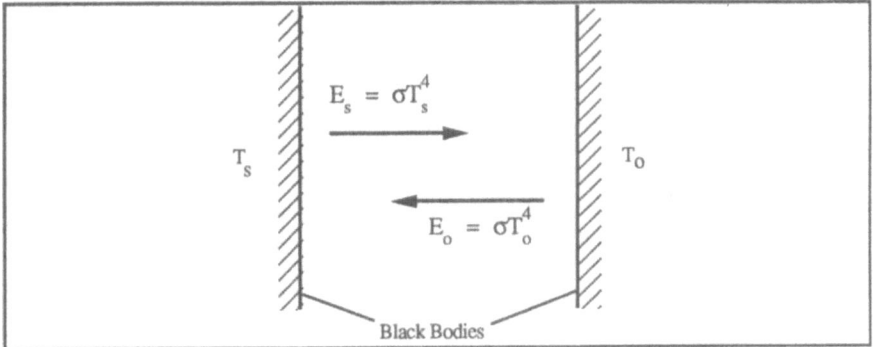

FIGURE 4.23   Radiative Energy Exchange Between Two Black Bodies

4.6 Numerical Methods in Heat Transfer and Fluid Mechanics

The laser machining process usually involves all three modes of heat transfer (conduction, convection and radiation) along with material phase change. Also, the erosion front often has a complicated three-dimensional geometry. These complications are difficult to model accurately using analytical techniques. Although experimental methods are available, they are expensive and time-consuming. An alternative to analytical and experimental methods is a numerical method, where the workpiece is discretized into a finite set of points or regions and the heat transfers between adjacent regions are analyzed in an iterative fashion. The resulting solution produces a temperature value for each point or region, and heat flows can be calculated by examining the temperature differences between adjacent points or regions. The advantage of using numerical methods is that a complicated geometry can be divided into smaller regions with simple geometries. Also, the accuracy and resolution of the solution depends on the size and number of regions used. The disadvantage of numerical methods is that highly-detailed models may require a significant amount of computational effort to evaluate. In developing a numerical model, a primary consideration is the trade-off between model detail and computational effort.

4.6.1 Basic Concepts

The numerical solution procedure depends on the mathematical type of the governing equation. It is convenient to consider the case of a general two-dimensional second-order equation, given as

$$A\frac{\partial^2 \phi}{\partial x^2} + B\frac{\partial^2 \phi}{\partial x \partial y} + C\frac{\partial^2 \phi}{\partial y^2} + D\frac{\partial \phi}{\partial x} + E\frac{\partial \phi}{\partial y} + F\phi + G = 0$$

$$(4\text{-}214)$$

where the coefficients depend on the independent variables and the dependent variable. A very important role is played by the sign of the discriminant $B^2 - 4AC$. Eq. (4-214) is an *elliptic* type when $B^2 - 4AC < 0$, *parabolic* when $B^2 - 4AC = 0$, and *hyperbolic* when $B^2 - 4AC > 0$.

Steady-state two-dimensional conduction problems are governed by the Laplace or Poisson equations. The temperature distribution for steady-state, constant-property conduction is governed by the Laplace equation. The Poisson equation governs steady-state conduction in the presence of heat generation. These equations are of the elliptic type and demand a specification of the boundary conditions at all the boundaries. The general forms of the Laplace and Poisson equations are written respectively as

$$\frac{\partial^2 \phi}{\partial x^2} + \frac{\partial^2 \phi}{\partial y^2} = 0$$

(4-215)

$$\frac{\partial^2 \phi}{\partial x^2} + \frac{\partial^2 \phi}{\partial y^2} + G = 0$$

(4-216)

where ϕ is a dependent variable, and G is a source term. In heat transfer problems ϕ becomes temperature and G is the rate of internal heat generation.

Transient conduction is governed by the following equation

$$\frac{\partial \phi}{\partial t} = A \left(\frac{\partial^2 \phi}{\partial x^2} + \frac{\partial^2 \phi}{\partial y^2} \right) + G$$

(4-217)

where A is the thermal diffusivity. Eq. (4-217) is parabolic in time and elliptic in space. An initial condition and two boundary conditions in each spatial coordinate are required. Eq. (4-217) can be solved numerically by starting with an initial condition and incrementing by discrete time intervals.

Fluid flow problems generally have a non-linear term due to the inertia or acceleration component in the momentum equation. In addition, heat transfer in a region may involve energy transport through motion of the medium along with conduction. For transient two-dimensional problems, the equations are of the form

$$\frac{\partial \phi}{\partial \tau} + u \frac{\partial \phi}{\partial x} + v \frac{\partial \phi}{\partial y} = A \left(\frac{\partial^2 \phi}{\partial x^2} + \frac{\partial^2 \phi}{\partial y^2} \right) + G$$

(4-218)

where ϕ denotes momentum, temperature, or some other transport variable, u and v are velocity components, A is the diffusivity for momentum or heat, and G is a forcing or heat source term. Eq. (4-218) is usually parabolic in time and elliptic in space. However, for the case of high speed flows, the terms on the left side dominate and the equation becomes hyperbolic in time and space. In this case, the hyperbolic equations can be solved by marching in time along certain characteristic directions.

The formulation of a problem requires a complete specification of the problem geometry and boundary or initial conditions. Spatial boundary conditions are of three general types expressed as

$$f = f(x)$$

(4-219)

$$\frac{\partial \phi}{\partial n} = f(x)$$

(4-220)

$$a(x)\phi + b(x)\frac{\partial \phi}{\partial n} = f(x)$$

(4-221)

where a and b are arbitrary functions of the independent variable x. The boundary conditions on ϕ in Eqs. (4-219-221) are often referred to respectively as Dirichlet, Neumann, and mixed boundary conditions.

In heat transfer problems, Dirichlet or Neumann boundary conditions arise when the temperature or the heat flux respectively are prescribed at a boundary. Mixed boundary conditions arise when convective heat is transferred to a conducting solid. For the case of fluid flow with heat transfer, the boundary conditions on the flow variables are usually expressed in terms of fluid velocities or shear stresses. Such boundary conditions are of the Dirichlet or Neumann type.

Numerical solutions for heat transfer and fluid mechanics problems are based on either the differential or integral formulation of the governing equation. Finite difference procedures have evolved as local or pointwise approximation methods for differential equations. Finite element methods have been developed from integral formulations applied over small regions of space surrounding one or more grid points. Although they may differ in mathematical formulation, both methods are designed to seek approximate solutions to governing equations for which closed-form results are unobtainable.

4.6.2   Finite Difference Method

The finite difference method transforms differential equations into a set of algebraic equations in order to solve for the dependent variable, with information from a continuum approximated by information at discrete nodal points. In the finite difference method, the continuous domain is replaced by a network or mesh of discrete points. Interactions between nodes on the network are modelled as algebraic equations. The algebraic equations are solved simultaneously to determine values for the dependent variable at all nodes.

A two dimensional region (Fig. 4.24) can be divided by grid lines spaced at uniform intervals Δx and Δy. The grid lines are indexed in two dimensions by integers i and j which increase monotonically along the x and y coordinates, respectively. A node exists at the intersection between any two grid lines, and the dependent variable $\phi(x,y)$ is defined at each node. In a general time dependent situation, the dependent variable $\phi(x,y,t)$ can be denoted by $\phi_{i,j}{}^{(m)}$, where the index m is the time interval. The time variable is incremented in discrete time steps Δt and is indexed such that $t_{m+1} = t_m + \Delta t$.

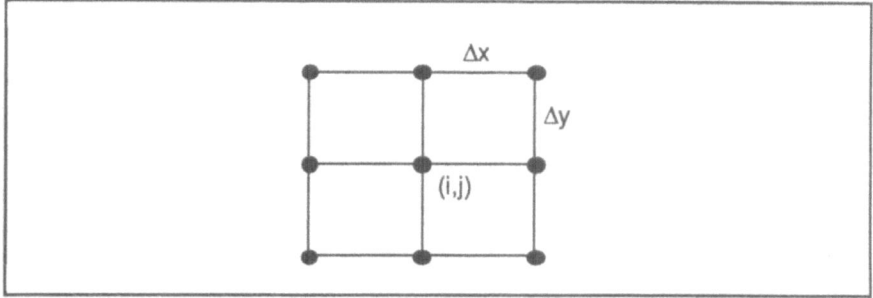

FIGURE 4.24   Two-Dimensional Grid Lines

As interactions between nodes occur, the dependent variable ϕ will vary along the grid directions. A possible variation of ϕ with x is sketched in Figure 4.25. This variation may be either time-dependent or steady-state. In order to transform a differential equation into an algebraic equation, the derivatives of ϕ with respect to x may be approximated in terms of discrete differences. For example, $\partial\phi/\partial x$ at (i,j) can be approximated by $\Delta\phi/\Delta x$, where the Δ denote discrete differences. The partial derivatives can be approximated as finite differences by using a Taylor series expansion. Three approximations for $\partial\phi/\partial x$ can be derived in several ways:

$$\left(\frac{\partial\phi}{\partial x}\right)_{i,j} \approx \frac{\phi_{i+1,j} - \phi_{i,j}}{\Delta x} = \delta_x^+ \phi_{i,j}$$

(4-222)

$$\left(\frac{\partial\phi}{\partial x}\right)_{i,j} \approx \frac{\phi_{i,j} - \phi_{i-1,j}}{\Delta x} = \delta_x^- \phi_{i,j}$$

(4-223)

$$\left(\frac{\partial\phi}{\partial x}\right)_{i,j} \approx \frac{\phi_{i+1,j} - \phi_{i-1,j}}{2\Delta x} = \delta_x \phi_{i,j}$$

(4-224)

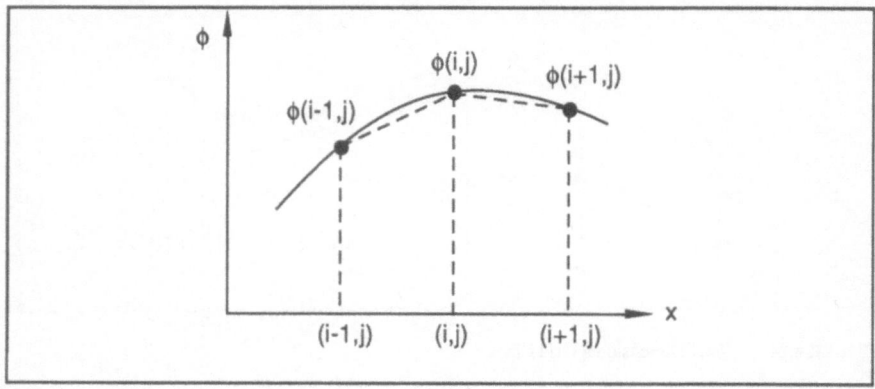

FIGURE 4.25   Variation of ϕ with x

The δ's are the finite difference operators. The three are forward (Eq. (4-222)), backward (Eq. (4-223)), and central (Eq. (4-224)) difference operators. Instead of deriving the tangent slope of the curve at a point as in the case of differentiation, these approximations employ the secant slopes to the right, to the left, and centered on the node point (i,j). The second derivative $\partial^2\phi/\partial x^2$ at (i,j) can be approximated by

$$\frac{\partial^2\phi}{\partial x^2} = \frac{\partial}{\partial x}\left(\frac{\partial\phi}{\partial x}\right) \approx \frac{\phi_{i+1,j} - 2\phi_{i,j} + \phi_{i-1,j}}{(\Delta x)^2} = \delta_x^2\phi_{i,j}$$

(4-225)

which is the central second difference approximation, denoted by the operator δ_x^2.

Steady state diffusion equations are of the Laplace or Poisson equations. For two-dimensional Cartesian coordinate systems with constant properties, the steady state diffusion equation applies

$$\frac{\partial^2\phi}{\partial x^2} + \frac{\partial^2\phi}{\partial y^2} + \frac{\dot{Q}}{k} = 0$$

(4-226)

Using the direct approach, suitable finite differences are directly substituted for partial derivatives in the governing equations. For the grid point (i,j) using the central difference approximations, the resulting approximation becomes

$$\frac{\phi_{i+1,j} - 2\phi_{i,j} + \phi_{i-1,j}}{(\Delta x)^2} + \frac{\phi_{i,j+1} - 2\phi_{i,j} + \phi_{i,j-1}}{(\Delta y)^2} + \frac{\dot{Q}_{i,j}}{k} = 0$$

(4-227)

For each interior point, an algebraic equation of the form given by Eq. (4-227) results when central second differences are used to approximate the Poisson equation. Special forms are required at irregular boundaries and at boundaries where heat flux conditions are prescribed.

The system of simultaneous equation may be written in a general matrix form as

$$(A)(\phi) = (F) \tag{4-228}$$

where (A) is a square matrix, (ϕ) is a column vector containing the unknown nodal values of ϕ, and (F) is column vector containing the source term and the prescribed boundary conditions. For example, when ϕ or its derivative are given at a boundary, terms containing these quantities are placed in (F). When mixed boundary conditions are prescribed, the nonhomogeneous terms that arise are placed in (F). (F) is known as the forcing vector.

For one-dimensional diffusion, the matrix (A) is tridiagonal, i.e., the only nonzero terms arise at the diagonal and on either side of it. In two and three dimensions, (A) has five and seven entries per row, respectively, but is usually not diagonal. From linear algebra, we know that unique solutions of Eq. (4-228) exist when the determinant of (A) is nonzero. This is usually the case in practical problems.

Iterative methods do not attempt to solve Eq. (4-228) directly. Instead such methods work with the individual algebraic equations, and do not require (A) and (F) to be constructed. In two dimensions, Eq. (4-227) reduces to

$$\frac{\phi_{i+1,j} + \phi_{i-1,j}}{(\Delta x)^2} + \frac{\phi_{i,j+1} + \phi_{i,j-1}}{(\Delta y)^2} - \left(\frac{1}{(\Delta x)^2} + \frac{1}{(\Delta y)^2}\right) 2\phi_{i,j} + \frac{\dot{Q}_{i,j}}{k} = 0 \tag{4-229}$$

Iterative methods start from an assumed initial field, and sequentially improve the field by using successive iterations until Eq. (4-229) is satisfied. With modern computers, the iterating equations are applied to all (i,j) values for which ϕ is unknown. Earlier manual methods applied the iterating equations selectively.

There are several iterative methods: Jacobi, the Gauss-Seidel, and relaxation methods. Iteration can be enhanced by extrapolating the Gauss-Seidel improvement. This is called relaxation, defined by

$$\phi_{i,j}^{n+1} = \frac{\omega}{\left(\frac{2}{(\Delta x)^2} + \frac{2}{(\Delta y)^2}\right)} \left(\frac{\phi_{i+1,j}^{n} + \phi_{i-1,j}^{n+1}}{(\Delta x)^2} + \frac{\phi_{i,j+1}^{n} + \phi_{i,j-1}^{n+1}}{(\Delta y)^2} + \frac{\dot{Q}_{i,j}}{k}\right) + (1 - \omega)\, \phi_{i,j}^{n} \tag{4-230}$$

where ω is called the relaxation factor. Iterative convergence is achieved with $0 < \omega < 2$; $\omega = 1$ corresponds to Gauss-Seidel iteration; $0 < \omega < 1$ to successive under-relaxation; and $1 < \omega < 2$ to successive over-relaxation.

The governing equation for two-dimensional transient diffusion is

$$\frac{1}{\alpha}\frac{\partial\phi}{\partial t} = \frac{\partial^2\phi}{\partial x^2} + \frac{\partial^2\phi}{\partial y^2} + \frac{\dot{Q}}{k}$$

$$(4\text{-}231)$$

A family of finite difference approximations may be constructed by using δ_x^2, and δ_y^2 for the spatial derivatives and a forward difference for the time derivative. By interpolating in time between levels m and m + 1, the resulting difference approximation of Eq. (4-231) at grid point (i,j) is

$$\frac{\phi_{i,j}^{m+1} - \phi_{i,j}^{m}}{\alpha\Delta t} = \gamma\left(\delta_x^2 + \delta_y^2\right)\phi_{i,j}^{m+1} + (1 - \gamma)\left(\delta_x^2 + \delta_y^2\right)\phi_{i,j}^{m} + \frac{\dot{Q}_{i,j}}{k}$$

$$(4\text{-}232)$$

where γ is a weighting factor in the range $0 \leq \gamma \leq 1$.

Eq. (4-232) applies at each nodal point for which the dependent variable ϕ is to be determined. The expression is modified near the boundaries to incorporate boundary conditions. If the nodal values of ϕ are known at time level m as initial data or as the result of a previous time advancement, Eq. (4-232) provides a set of simultaneous equations for the unknown ϕ's at m + 1. Solution of this set yields $\phi_{i,j}^{(m+1)}$. This process may then be repeated, allowing the ϕ field to be marched in time.

The procedures used to solve the equation will depend on the value of the weighting factor. When $0 < \gamma \leq 1$, Eq. (4-232) contains five values of ϕ at level m + 1. The resulting equations must be solved simultaneously and the system is implicit. Both iterative and direct solution procedures may be employed. However, in contrast with steady state problems, the simultaneous equations must be solved at every time advancement. When $\gamma = 1/2$, the finite difference approximation is centered in time at level m + 1/2, and the time discretization is second order correct. This is the Crank-Nicolson method. When $\gamma = 1$, the approximation is centered on grid point (i, j, m + 1). The spatial derivatives are evaluated at level m + 1. This is a fully implicit method.

In matrix form, the set of equations represented by Eq. (4-232) is

$$(A)\ (\phi^{(m+1)}) = (B)\ (\phi^{(m)}) + (F) \qquad\qquad (4\text{-}233)$$

The unknown nodal values of ϕ appear on the left side and (A) and (B) contain physical properties and mesh increments. The forcing vector (F) may be time dependent through the source term or the boundary conditions. When $(A)^{-1}$ exists, we have

$$(\phi^{(m+1)}) = (C)\ (\gamma\phi^{(m)}) + (A)^{-1}\ (F) \qquad\qquad (4\text{-}234)$$

where $(C) = (A)^{-1}\ (B)$

There are several properties of numerical solutions which determine the level of accuracy and detail that can be achieved. These properties are *stability*, *consistency*, and *convergence.* Stability of a numerical solution describes whether or not the dependent variable is bounded. For a steady-state solution, the dependent variable is unstable if oscillation occurs. For a transient analysis, the dependent variable is unstable if the solution oscillates with an amplitude that increases with time. For second-order differential equations such as those encountered in heat transfer and fluid mechanics, instability is usually caused by discretization of the first-order derivative with an improper time interval.

Consistency is a property based on the errors resulting from truncation of numbers during numerical computation. A discretized equation is consistent if the error between the numerical and analytical solutions approach zero as the time and spatial increments approach zero. Truncation errors primarily occur during application of Taylor's expansion of the terms in the differential equation. In numerical methods, only the first few terms of the Taylor's expansion are used. Since Taylor's expansion is an infinite series, the residual higher-ordered terms represent the truncation error between the approximate and exact solutions. There are two ways to reduce the truncation error. First, the mesh size can be decreased; however, this reduction may increase the round-off error during computation and will always increase the computation time and effort. Second, a proper mesh size ratio can be chosen such that the effects from higher-order terms are cancelled. In transient conduction analysis, the mesh size ratio is called the Fourier number (Fo), and the stability criterion is expressed as:

$$Fo = \frac{\alpha \, \Delta t}{\Delta x^2} < \frac{1}{6}$$

(4-235)

for the case of cenvection, the stability criterion is:

$$\Delta t < 0.5 \frac{\Delta x}{V_\infty}$$

(4-236)

The third property of numerical methods is the degree of convergence of the solution. In order to yield meaningful results, the dependent variable solution should be bounded at all times. A numerical solution is said to be convergent if the discretization error approaches zero as the mesh is refined. The convergence criterion can be mathematically expressed as:

$$\frac{\left| \phi_{i+1,j}^n - \phi_{i,j}^n \right|}{\phi_{i,j}^n} \to 0 \quad \text{and} \quad \frac{\left| \phi_{i,j+1}^n - \phi_{i,j}^n \right|}{\phi_{i,j}^n} \to 0 \qquad \text{as i and j} \to \infty$$

(4-237)

Convergence means that the exact solution is approached numerically through mesh refinement by either increasing the number of spatial or time steps. For finite difference solutions, the Lax theorem [16] states that the convergence of a solution is ensured if and only if the solution is both stable and consistent.

4.6.3   Finite Element Methods

Finite element methods provide piecewise, or regional, approximations to partial differential equations, while finite difference methods generally provide pointwise approximations. When irregular geometries or unusual boundary conditions are present, it may be desirable to use finite element methods instead of finite difference methods. A finite element is a discrete spatial region that is a subdivision of a continuum. Many geometric shapes for finite elements can be used which will completely fill the interior and approximate the contours of the bounding surface S. A common two-dimensional element is the triangle, and a three-dimensional element is the tetrahedron. With such elements, it is possible to approximate many complex and irregular objects.

The finite element methods are established on the basis of integral minimization. Early formulations used the calculus of variations to replace a differential equation by a maximized or minimized regional integral. However, for transient heat diffusion or convection heat transfer, it is necessary to use general procedures such as a method of weighted residuals rather than variational calculus. Recent formulations are based on weighted integrals of a differential equation over a region of space.

Consider the geometry in Figure 4.26, which illustrates a volume V with a bounding surface S. In any part of V, the time dependent diffusion equation can be applied. The diffusion equation may be integrated over the volume V, or any subvolume V', to obtain

$$\int_V \left(\rho c_p \frac{\partial \phi}{\partial \tau} - \nabla \cdot (k \nabla \phi) - \dot{Q} \right) dV = 0$$

$$(4\text{-}238)$$

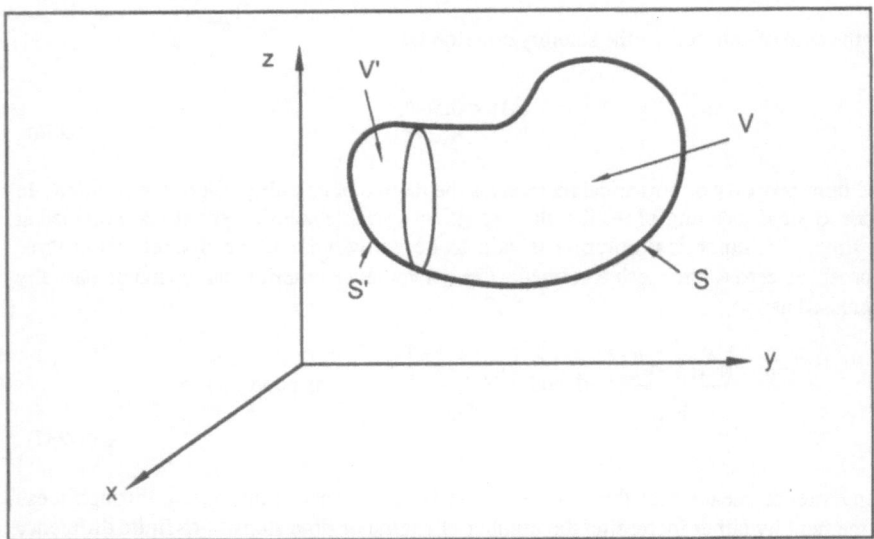

FIGURE 4.26   Geometry for Finite Element Analysis

Eq. (4-238) formed the basis for the control volume finite difference approach, where certain volume integrals were converted to surface integrals. In turn, different profile assumptions were used to evaluate the surface integrals and any remaining volume integrals. By way of contrast, a finite element approach does not simply discretize surface integrals, nor does it employ different profile assumptions for volume and surface integrals. Instead, a consistent representation of the ϕ field is proposed and is used to evaluate all the integrals in Eq. (4-238).

The finite element method involves the following logical steps.

1. The region is subdivided into elements.

2. Interpolation or shape functions are selected for the elements. The spatial solution can be expressed as a linear combination of all interpolation functions.

3. The matrix equations for an individual element are formulated using the integral statement for the element as a guide.

4. The matrix equations for the overall system, consisting of all the elements, are assembled. The global equations are of the same form as the element equations but are of larger dimension.

5. The global equations are solved.

The next step is to postulate a spatial form for the dependent variable ϕ. This will allow evaluation of the spatial integrals in Eq. (4-238). The unknown function ϕ is replaced by an approximate function, where x denotes the position vector. In turn, $\phi(x,t)$ is expanded as a spatially weighted summation over all the nodal points N. This may be expressed as

$$\phi(\bar{x},\tau) \approx \hat{\phi}(\bar{x},\tau) = \sum_{i=1}^{N} p_i(\bar{x})\, \phi_i(\tau)$$

(4-239)

In Eq. (4-239), the $\phi_i(t)$ are quantities defined at the node points. They represent the unknown nodal amplitudes of the problem. The $p_i(x)$ define the spatial form of the solution between node points and are called the shape or interpolation functions. Each term in the summation represents the contribution from a particular node i and contains a quantity ϕ_i.

The shape functions depend on the type of elements used and the number of nodes in each element. Generally, the shape functions are simple analytical functions (i.e., polynomials) that have some similarity to the expected form of the solution. For example, if simple continuity of ϕ is imposed for the three node planar triangles shown in Figure 4.27, the appropriate interpolating functions are linear polynomials. $p_i(x)$ varies from unity at node i to zero at other nodes. The nodal amplitudes $f_i(t)$ are interpolated linearly between nodes, and the composite function ϕ varies linearly over the triangles.

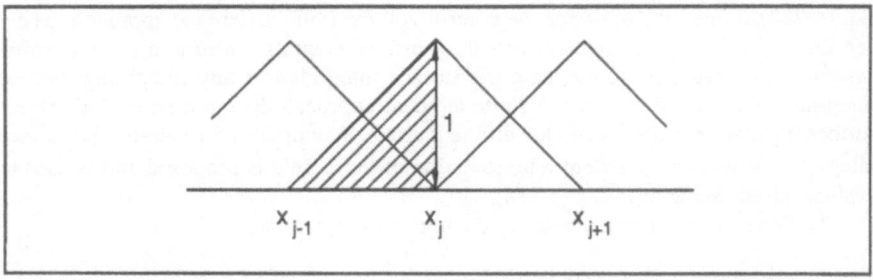

FIGURE 4.27   Linear Interpolation Functions

Eq. (4-238) is applied to any finite element, or to a region. Substituting $\phi(x,t)$ into the integrand of Eq. (4-238) does not, in general, make the integrand zero. If it did, ϕ would be a solution of the partial differential equation. Instead, the substitution leads to a residual ε, defined by

$$\rho c_p \frac{\partial \bar{\phi}}{\partial t} - \nabla \cdot (k \nabla \bar{\phi}) - \dot{\bar{Q}} = \bar{\varepsilon}$$

(4-240)

The residual represents the error incurred by approximating ϕ with ϕ- in the governing differential equation. Substituting Eq. (4-240) into Eq. (4-239) the following is obtained:

$$\int_V \bar{\varepsilon} \, dV = 0$$

(4-241)

Applying this equality to each element leads to the requirement that the average error in each element be zero. Eq. (4-241) forms the basis of a control volume finite element approach and represents a special case of the method of weighted residuals.

The method of weighted residuals is a general procedure for obtaining approximate solutions of linear and non-linear partial differential equations. The approach involves choosing N linearly independent weighting functions W and setting each weighted integral equal to zero over the entire solution domain.

$$\int_V W_i \bar{\varepsilon} \, dV = 0 \quad i = 1,2,\ldots,N$$

(4-242)

Eq. (4-242) represents a set of N simultaneous equations. A solution of the set implies that the residual error is zero in the sense of a weighted average.

By choosing interpolation functions p_i that are nonzero only in the immediate neighborhood of a node i, the governing equations provide the element equations for every element.

References

1.   Ames, W.F., *Numerical Methods for Partial Differential Equations*, Barnes and Noble, New York, 1977.

2.   Arpaci, V.S., *Conduction Heat Transfer*, Addison-Wesley, Cambridge, MA, 1966.

3.   Carslaw, H.S., and J.C. Jaegar, *Conduction of Heat in Solids*, Clarendon Press, Oxford, England, 1959.

4.   Donaldson, C.D., and R.S. Snedeker, "A Study of Free Jet Impingement. Part 1. Mean Properties of Free and Impinging Jets," *Journal of Fluid Mechanics,* Vol. 45 (1971), 281-319.

5.   Eckert, E.R.G., and R.M. Drake, *Heat and Mass Transfer*, McGraw-Hill, New York, 1959.

6.   Jaluria, Y., and K.E. Torrance, *Computational Heat Transfer*, Hemisphere Pub. Co., Washington D.C., 1986.

7.   Landau, L.D., and E.M. Lifshitz, *Mechanics of Continuous Media*, Addison-Wesley, Cambridge, MA, 1959.

8.   Liepmann, H., and A. Roshiko, *Elements of Gas Dynamics*, J. Wiley and Sons, New York, 1956.

9.   Love, E.S., C.E. Grisby, L.P. Lee, and M.J. Woodling, "Experimental and Theoretical Studies of Axisymmetric Free Jets," NASA TR R-6.

10.   Ozisik, M.N., *Boundary Value Problems of Heat Conduction*, International Textbook Co., Scranton, PA, 1968.

11.   Pitts, D. R. and L. E. Sissom, *Theory and Problems of Heat Transfer*, McGraw-Hill, New York, 1977.

12.   Prandtl, D, *Essentials of Fluid Dynamics*, Hafner, New York, 1952, p. 266.

13.   Rosenthal, D., "The Theory of Moving Sources of Heat and Its Applications to Metal Treatments," `Transactions of the ASME* (Nov. 1946).

14.   Schneider, P.J., *Conduction Heat Transfer*, Addison-Wesley, Cambridge, MA, 1955.

15. Shapiro, A.H., *Compressible Fluid Flow, Vol. 1*, Donald Press, New York, 1953.

16. Shih, T.-M., *Numerical Heat Transfer*, Hemisphere Pub. Co., Washington D.C., 1984.

17. Voller, V.R., M. Cross, and N.C. Markatos, "An Enthalpy Method for Convection/Diffusion Phase Change," *International Journal of Numerical Methods in Engineering*, Vol.24 (1987), 271-284.

5
Laser Machining Analysis

This chapter surveys many of the most important theoretical works in laser surface heating, drilling, cutting, grooving and three-dimensional machining found in recent literature which are based on an understanding of the physics of laser/material interaction. These process models are needed in order to choose correct operating parameters and to implement closed-loop process control. These works entail both analytical and numerical modelling to find relationships between operating parameters, temperature distribution and erosion front geometry. The phenomena occurring during laser machining processes, such as plasma formation, creation of striations, and changes in surface absorption of laser beam energy are also explained in a theoretical context.

5.1 Introduction

The modelling of a manufacturing process, or the mathematical description of the physical phenomena occurring during the process, is a critical element for its successful operation. However, process models can contribute a great deal to the *optimization* and *control* of the manufacturing process. Creating mathematical descriptions of manufacturing processes is a complex task, since these processes usually involve a number of physical mechanisms, some of which are more dominant than others.

All manufacturing processes, including laser machining, have operating parameters or setting variables which determine the operation of the process. Setting variables for laser machining include the laser power, the spot size of the focused laser beam, the speed of the workpiece, etc. To optimize laser machining, one must select setting variables so that the process optimizes one or more performance measures such as material removal rate or surface quality. To arrive at optimum setting variables, experiments can be made with a variety of setting variables to find the optimum conditions based on the results, or a process model can be created and manipulated to lead to the selection of setting variables which will optimize the desired performance measures.

While the selection of optimum setting variables prior to the actual execution of the process can lead to optimum performance of the process, in reality this is only the case if there are no disturbances in the process. In laser machining, such disturbances can include fluctuations in laser beam power, changes in focal spot size, material property variations, and a host of other unpredictable factors that can lead to substantial variation of the outcome compared to what had originally been predicted by the selection of the setting variables. This introduces the need for *process control*, namely for continuous or periodical variation of the setting variables so that the outcome of the process is as desired. Such a control scheme for laser machining (Fig. 5.1) requires sensor models, or relationships between measured variables such as acoustic emission, temperature and

160

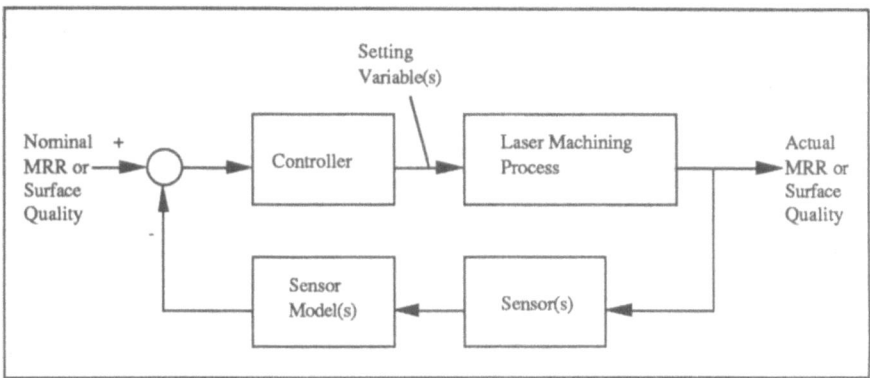

FIGURE 5.1   Process Control Scheme for Laser Machining

optical measurements, with state variables such as depth of cut or hole taper. It also requires a control strategy which involves a model of the laser machining process that relates state variables (Fig. 5.1) and their variations to the setting variables. Hence, a mathematical description of the process is critical for the successful application of any process control procedure.

Any laser machining process involves the formation of an *erosion front* (Fig. 5.2), where the interaction of the laser beam with the workpiece material results in material removal. The power of the laser beam, the workpiece material and a number of other factors influence the shape and the propagation velocity of the erosion front (Fig. 5.3). When the laser beam radiates the surface of the workpiece, the material is heated up due to photon absorption, and if the power density of the incident beam is high enough, a phase change occurs, turning the solid material to a liquid or vapor. The zone that has undergone phase transformation and has heated up acts as a heat source itself, and transfers heat to the rest of the material and to the environment through conduction, convection and/or radiation. In cases of high power density, the material may be vaporized, and further energy absorption by the vapor may create a plasma formation which acts both as an energy absorbant and a heat source. This plasma may cover the erosion front, partially shielding it from further laser radiation, while the material under the surface skin heats up and may melt and vaporize, creating an explosion under the surface which removes material. Depending on the specific process and material, these experimentally observed phenomena are either dominant or insignificant; however, all laser machining processes involve heat transfer.

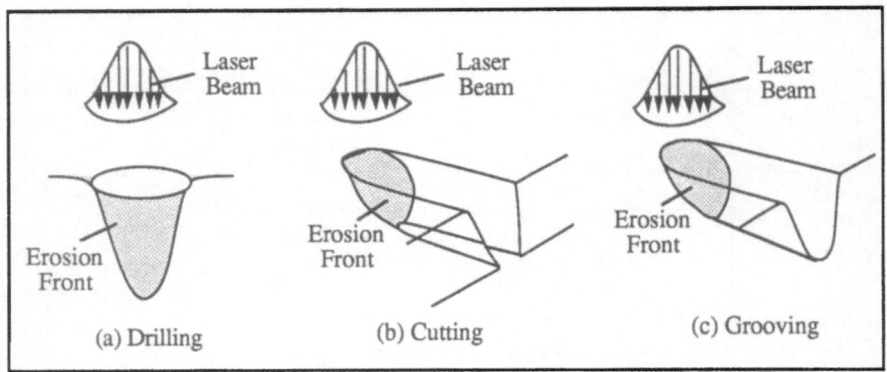

FIGURE 5.2   Schematics of Drilling, Cutting, and Grooving Processes

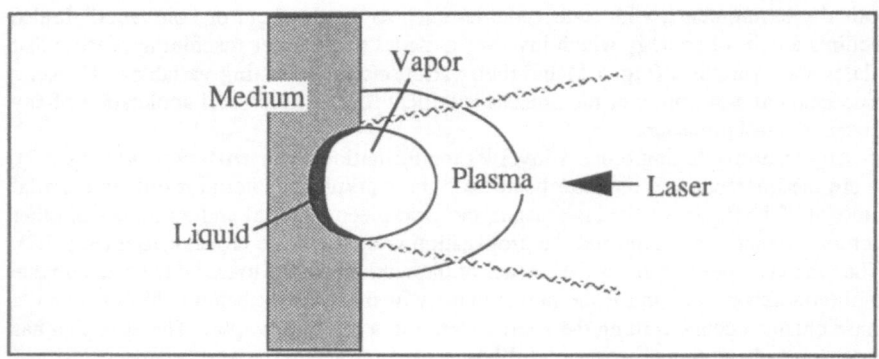

FIGURE 5.3   Laser Beam Interaction With a Medium

5.2 Problem Definitions for Laser Process Modelling

Examples of processes that involve material removal are drilling, cutting, and grooving. These processes are different from welding in that material is removed from a workpiece in liquid and/or vapor form. Drilling is by nature a non-steady process. Until a workpiece surface reaches the phase transition temperature, material removal does not occur. When material is removed, the temperature distribution inside the material and hole shape change continuously with respect to coordinates fixed at the laser beam. In cutting and grooving, the kerf shape and temperature distribution do not change with respect to a coordinate system fixed at a laser beam; therefore cutting and grooving can be treated as steady processes.

5.2.1   Drilling

In laser drilling, the motion of the erosion front and temperature distribution with time has a significant effect on the surface absorption of beam energy. The absorptivity, along with beam intensity, was found to be the most important factors in the drilling process in [40]. Absorptivity and/or reflectivity variation during laser drilling was investigated in [54, 55]. For laser power densities below a threshold value ($300J/cm^2$ for copper), the surface for most metals shows a high reflectivity to beam energy and no material removal occurs. During the initial stages of beam/material interaction, the surface reflectivity is time-dependent, since the slope of the hole wall changes rapidly with time. During this stage, a significant portion of the incident beam energy is reflected from the erosion front. As time progresses, absorptivity increases and reaches a final value, since the slope of a developed hole profile does not change significantly with time.

For energy densities above the threshold, drilling occurs but may be hampered by the formation of plasma plumes or laser supported detonation (LSD) waves. The sequential steps of the plume evolution caused by a ruby laser pulse was investigated [40]. LSD waves occur when the plasma formed in the hole couples with a shock wave. The shock wave provides a sufficient number of electrons to initiate a breakdown of the gas behind the shock wave, which causes absorption of all laser radiation by a mechanism to break down the coherent waves. Therefore, effective laser drilling depends on quick dissipation of LSD waves to minimize their effects. In [54], a model relating the fraction of laser energy required to create a crater of given dimensions is given as:

$$\alpha = y\left(xQ_v + (1-x)Q_L\right)$$

(5-1)

where α is the fraction of beam energy needed to create a crater of given volume, Q_v and Q_L are the vaporization and melting energies per unit volume of material, y is the ratio of energy required to melt or vaporize the crater volume to pulse energy, and x is an estimate of the fraction of crater material vaporized versus melted. A utilization factor for laser energy can be expressed as:

$$\eta = \frac{1}{E_p} \int_{t_0}^{\infty} P(t)\, dt$$

(5-2)

where η is the utilization factor, E_p is the pulse energy, P(t) is the beam power and t_0 is the time when the LSD wave is dissipated. The utilization factor shows the fraction of beam energy which is actually engaging the erosion front. Efficient laser drilling is related to minimizing t_0. One method of reducing the LSD propagation time is by using a small F-number or small focal length focussing lens. This allows a smaller focussed spot and higher beam intensities to be achieved. Another method is to use a low density gas in place of air for the coaxial jet.

Drilling is usually performed with a pulsed laser, which produces a higher intensity and evaporates the material easily. The effects of a varying length of pulse on laser drilling were investigated [28]. Pulsed laser beams create a large temperature gradient inside the material because of their high intensity and short interaction time. This causes a large thermal stress which induces cracks in the material. Moreover, in drilling a material is initially at a constant temperature. The temperature profile and the tangential stress distribution of the laser-formed hole were calculated to qualitatively judge the potential for thermal fracture in alumina material when heated by a laser beam [40]. During drilling, residual stresses were measured from strain measurement near a hole [6]. Drilling was performed over various materials [2, 32]. Hole depth and shape were predicted as a function of absorbed intensity [57]. Material removal during laser drilling was also investigated. Material is removed in the form of liquid and/or vapor. The driving force for the material expulsion was known to be the high pressure due to the explosive evaporation. Also the expulsion mechanism of liquid material during drilling was studied in [40, 54, 57].

In order to obtain a useful model and still solve the problem, the following assumptions need to be made [19]:

1. the laser beam intensity I is sufficient to cause vaporization of the front surface of the material
2. the gas created by the vaporization of the material is transparent to the incident laser energy
3. heat losses due to radiation are negligible
4. the thermal constants and optical absorption coefficient b are independent of the laser beam intensity and the temperature of the solid
5. the effects of radial heat conduction and the liquid phase can be ignored. This is generally true for materials in which the thermal penetration depth is less than the beam diameter.

The energy balance at the front surface requires that the energy is conducted from the solid. Thus,

$$\rho L_v \dot{Z} = k \left(\frac{\partial T}{\partial z} \right)_{z=Z}$$

(5-3)

where T is the temperature, k is the thermal conductivity, ρ is the density of the solid, L_v is the heat of vaporization, and Z is the depth to which material has been removed. The velocity of the front surface, the time differential of Z is $\dot{Z}$.

In order to determine the temperature profile within the material, the heat conduction equation with a distributed laser heat source moving with the front surface is used. Thus,

$$\rho c_p \frac{\partial T}{\partial t} = bI \, e^{-b(z-Z)} + k \frac{\partial^2 T}{\partial z^2}, \quad z > Z$$

(5-4)

where c_p is the specific heat, b is the optical absorption coefficient, I is the beam intensity, and t is time.

Solving (5-3) and (5-4) with appropriate boundary conditions will yield the temperature and velocity. The appropriate boundary conditions are

$$T = T_V , \text{ at } z = Z \qquad (5-5)$$

and

$$T = T_\infty \text{ at } z = \infty \qquad (5-6)$$

Two initial conditions are required for (5-3) and (5-4). The first assumes that at time $t = 0$, the drilling process begins. Thus,

$$Z = 0 \text{ at } t = 0 \qquad (5-7)$$

The second initial condition depends upon how the material is heated before the drilling process has begun. Although it could be done using other techniques, the penetrating is generally done by earlier portions of the laser pulse. The solution of the heat equation with a distributed laser heat source could be written in terms of exponentials, so it would be convenient to write the last initial condition describing the temperature profile in the material at t=0 in terms of an exponential function. Instead of expanding the initial condition formally, the constant q is used as a fitted parameter that would most closely match the determined initial temperature profile. Thus at $t = 0$

$$T = (T_V - T_\infty)(1 + qz) e^{-qz} \qquad (5-8)$$

Equations (5-3) through (5-8) can be combined and nondimensionalized with respect to a moving reference frame so that

$$\frac{\partial \theta}{\partial \tau} - u \frac{\partial \theta}{\partial s} - \frac{\partial^2 \theta}{\partial s^2} = \frac{B}{\lambda} e^{-Bz} \qquad (5-9)$$

and

$$u = \lambda \left(\frac{\partial \theta}{\partial s} \right)_{s=0} \qquad (5-10)$$

subject to the following conditions:

$$\text{at } t = 0 \quad \theta = (1 + Qs) e^{-Qs} \qquad (5-11)$$
$$\text{at } s = 0 \quad \theta = 1 \qquad (5-12)$$
$$\text{at } s = \infty \quad \theta = 0 \qquad (5-13)$$

where

$$\theta = (T - T_\infty)/(T_V - T_\infty) \qquad (5\text{-}14)$$

$$s = (Ic_p/kL_v)\,(z - Z) \qquad (5\text{-}15)$$

$$u = (rL_v{}^2/I)\dot{Z} \qquad (5\text{-}16)$$

$$\tau = (I^2 c_p/\rho k L_v{}^2)\,t \qquad (5\text{-}17)$$

A heating parameter λ, absorption parameter B, and a normalized initial parameter Q are defined as follows:

$$\lambda = (c_p T_v)/L_v \qquad (5\text{-}18)$$

$$B = (kbL_v)/(Ic_p) \qquad (5\text{-}19)$$

$$Q = (kqL_v)/(Ic_p) \qquad (5\text{-}20)$$

Initially $(\partial\theta/\partial s)_{z=0}$ is zero. Thus the normalized inner velocity u is zero for small values of t. Substituting this value of u into Eq. (5-9) gives

$$\frac{\partial\theta_i}{\partial\tau} - \frac{\partial^2\theta_i}{\partial s^2} = \frac{B}{\lambda}e^{-Bs} \qquad (5\text{-}21)$$

where θ_i is the temperature based upon the initial velocity and is subject to the conditions given in Eq. (5-11-5.13). By taking the Laplace transform of Eq. (5-21) with respect to t, solving the resulting second order ordinary differential equation, and taking the inverse transform, a solution to Eq. (5-21) can be obtained. Knowing θ_i the dimensionless inner velocity can be recalculated as

$$u_i = \lambda\left(\frac{\partial\theta_i}{\partial s}\right)_{s=0} \qquad (5\text{-}22)$$

Therefore

$$u_i = 1 - e^{B^2\tau}\,\mathrm{erfc}(B\tau^{1/2}) + 2\lambda Q^3\tau\,e^{Q^2\tau}\,\mathrm{erfc}(Q\tau^{1/2}) - \frac{2\lambda}{\pi^{1/2}}\,Q^2\tau^{1/2} \qquad (5\text{-}23)$$

where erfc is the complementary error function. The drilling speed can be integrated with respect to time in order to derive a solution for hole depth. The temperature distribution during laser drilling can be calculated for the cases of infinite and finite workpiece thicknesses. For any instantaneous point source Q, the temperature field can be described by:

$$T(x,y,z,t) = \frac{Q(x',y',z')}{8(\pi K t)^{3/2}} \exp\left(-\frac{(x - x')^2 + (y - y')^2 + (z - z')^2}{4Kt}\right)$$

(5-24)

where K is the thermal diffusivity and x', y', and z' are the heat source coordinates. To simplify the solution procedure, the cartesian coordinate system can be transformed to a cylindrical coordinate system. Using the method of images [40], the temperature distribution for an infinite workpiece thickness can be calculated as:

$$T(r,z,t) = \int_0^a \int_0^t \frac{P(r',t')}{4\rho c_p \sqrt{\pi} \left(K(t - t')\right)^{3/2}} \exp\left(-\frac{r^2 + r'^2}{4K(t - t')}\right)$$

$$\cdot I_0\left(\frac{r r'}{2K(t - t')}\right) r' \left(\exp\left(-\frac{(z - f(t'))^2}{4K(t - t')}\right) + \exp\left(-\frac{(z + f(t'))^2}{4K(t - t')}\right)\right) dt' \ dr'$$

(5-25)

where P(r',t') is the beam intensity at coordinate r' from the beam center and time t' from beam activation, a is the hole radius (assumed to be equal to the beam radius), and f(t') is the hole depth as a function of time from solution of Eq. (5-25). For the case of a finite workpiece thickness, the temperature distribution is:

$$T(r,z,t) = \int_0^a \int_0^t \frac{P(r',t')}{2\rho c_p K(t - t')d} \exp\left(-\frac{r^2 + r'^2}{4K(t - t')}\right) I_0\left(\frac{r r'}{2K(t - t')}\right)$$

$$\left(1 + \sum_{n=1}^{\infty} \cos\left(\frac{n\pi(z + f(t'))}{d}\right) \exp\left(-\frac{Kn^2\pi^2(t - t')}{d^2}\right) + \sum_{n=1}^{\infty} \cos\left(\frac{n\pi(z - f(t'))}{d}\right) \exp\left(-\frac{Kn^2\pi^2(t - t')}{d^2}\right)\right)$$

$$r' \ dt' \ dr'$$

(5-26)

where d is the workpiece thickness.

Thermal stresses can be induced in the workpiece during laser drilling when regions cannot expand freely under heating. In the case of large temperature gradients, stresses which exceed the fracture limit of the material can build up. For the case of plane strain, the distribution of tangential stresses near the erosion front can be found as:

$$\sigma_\theta = \left(\frac{\alpha E}{1 - \nu}\right) \frac{1}{r^2} \left(\int_a^r T r \, dr - T r^2\right)$$

(5-27)

where σ_θ is the tangential stress, α is the coefficient of thermal expansion, E is the elastic modulus, and ν is Poisson's ratio. An alternative method for analyzing the laser drilling problem is through a numerical solution. In [57], a one-dimensional finite difference model was developed to determine hole depth and temperature distribution. The material is assumed to be semi-transparent; a portion of the beam energy not absorbed at the erosion front is transmitted into the workpiece. The power density of the beam as it penetrates into the workpiece has the form:

$$e(y) = (1 - R_\infty) e_0 \exp(-\alpha y)$$

(5-28)

where R_∞ is the surface reflectivity from the erosion front, e_0 is the focussed power density, and α is the absorption coefficient in the interior of the workpiece. This beam penetration effect can be modeled as an equivalent internal heat source w(y), where:

$$w(y) = -\frac{de(y)}{dy} = (1 - R_\infty) e_0 \alpha \exp(-\alpha y)$$

(5-29)

The workpiece can be divided into n nodes with a width of δ. Each node undergoes a heating sequence. First, the node absorbs heat at a constant specific heat until the melting temperature is reached. At the melting point, the node absorbs energy at a constant temperature until the heat of fusion has been reached. Once the heat of fusion is reached, the node absorbs energy at a constant specific heat until the vaporization temperature is reached. Finally, the node absorbs energy at the vaporization temperature until the heat of vaporization has been reached; then the node is effectively vaporized. For the ith node, the heat transfer can be modelled as:

$$\rho c_p \, \Delta T_i = \frac{Q_i}{\delta}$$

(5-30)

where ΔT is the temperature rise at the node for the time step Δt and Q_i is the heat input to the node, given as:

$$Q_i = \left(\frac{k}{\delta}(T_{i-1} - T_i) - \frac{k}{\delta}(T_i - T_{i+1}) + \delta w(y_i)\right) \Delta t$$

(5-31)

where k is the thermal conductivity and $y_i = (i - 1)\delta$. When the node reaches the melting temperature, an auxiliary variable P_i is defined. This variable accumulates the calculated value of heat input $c_p \Delta T_i$ for enough iterations to account for the heat of fusion. When the accumulated energy equals the heat of fusion, P_i is switched off and heat transfer continues according to Eq. (5-31). When the node reaches the vaporization temperature, the Pi variable is switched on again. When the energy accumulated in Pi is equal to the combined heats of fusion and vaporization, the node vaporizes and vanishes. If a total of m nodes have been vaporized, then $y_i = (i - 1 - m)\delta$. For nodes which have been vaporized, the heat input can be modified in the following manner:

For i = m+1, m+2,...,n-1

$$Q_i = \frac{k}{\delta}(T_{i-1} - T_i) - \frac{k}{\delta}(T_i - T_{i+1}) + \delta(1 - R_\infty)e_0\alpha \exp\left(-\alpha(i-1-m)\delta\right)$$

(5–32)

For i=n

$$Q_i = \frac{k}{\delta}(T_{i-1} - T_i) - \varepsilon\sigma(T_i^4 - T_\infty^4) + \delta(1 - R_\infty)e_0\alpha \exp\left(-\alpha(i-1-m)\delta\right)$$

(5-33)

where ε is the surface emissivity and σ is the Stefan-Boltzmann constant (given in Section 4.2.1).

5.2.2 Cutting

Laser cutting is the most widely-used of all laser machining processes. In laser cutting, a laser beam penetrates through the entire thickness of the workpiece and advances parallel to the surface of the workpiece. Depending on the phase of material removed, laser cutting is divided into two types: sublimation and fusion cutting. In the case of sublimation cutting, the material is vaporized; sublimation is usually achieved for materials with low vaporization temperatures and heats of vaporization (such as plastics) and/or by applying high energy densities generated from pulsed solid state lasers. For laser fusion cutting, the material is melted at the erosion front and ejected from the kerf with the help of an inert gas jet. If a reactive gas jet such as oxygen is used, the process is called a reactive gas assisted cutting. In this case, chemical reactions between the gas jet and the material serves as a significant secondary mechanism for material removal. In some instances where an oxygen jet is applied to laser cutting of steel, oxidation reactions may become the primary mode of material removal; the process becomes similar to plasma arc cutting.

One goal of laser cutting is to achieve the highest material removal rate or cutting speed possible for a given workpiece thickness. Instead of relying on experimental calibration of relationships between cutting depth to laser parameters and material properties, models can be used which describe the physics of the laser cutting process. In [44], a cutting model was developed based on the absorptivity calculated over the cutting kerf. In [5] a relationship between the power density incident on a material and the resulting cutting speed is developed in terms of the thermal properties of the material. The theory in [5] indicates that to maximize cutting speed and energy utilization, the jet diameter should be as small as possible consistent with an attainable power density.

In [41], a laser erosion front model was formulated based on a balance between absorbed beam power, power for melting material, power for heating material, and conduction heat losses. This model makes several assumptions. First, the kerf width is assumed to be constant at the laser beam diameter. Second, the contour lines of the erosion front can be described by semi-circles with radius equal to the beam radius. Third, the inclination angle of the erosion front is constant and determined experimentally. Finally, the erosion front is assumed to have a linear slope. The model determines the effect on the shape of the erosion front from beam power, beam polarization, laser mode, focal length, beam diameter, and focal position. The first step in this model is to determine the absorptivity of the erosion front. The refractive index and absorption coefficient can be described by:

$$
n = \left(0.5 \left(\left(\left(1 - \frac{\omega_p^2}{v^2 + \omega^2} \right)^2 + \left(\frac{v\,\omega_p^2}{\omega(v^2 + \omega^2)} \right)^2 \right)^{1/2} + \left(\frac{1 - \omega_p^2}{v^2 + \omega^2} \right) \right) \right)^{1/2}
$$

(5-34)

$$
\alpha = \left(0.5 \left(\left(\left(1 - \frac{\omega_p^2}{v^2 + \omega^2} \right)^2 + \left(\frac{v\,\omega_p^2}{\omega(v^2 + \omega^2)} \right)^2 \right)^{1/2} - \left(\frac{1 - \omega_p^2}{v^2 + \omega^2} \right) \right) \right)^{1/2}
$$

(5-35)

where ω is the laser frequency, ω_p is the plasma frequency, and v is the electron collision frequency. A complex refractive index can be defined as:

$$
n^* = n + i\,\alpha \frac{c_o}{2\omega}
$$

(5-36)

where c_o is the speed of light and $i = (-1)^{1/2}$. The absorptivity of the melt film at the erosion front can be expressed as:

$$
A_s = 1 - \left(\frac{\cos\delta - (n^{*2} - \sin^2\delta)^{1/2}}{\cos\delta + (n^{*2} - \sin^2\delta)^{1/2}} \right)^2
$$

(5-37)

where δ is the angle of incidence of the laser beam on the erosion front. If the mode, polarization and focal shape of the beam are known, then the variation of absorptivity on the erosion front can be determined as:

$$A(z,\beta)P_{Laser} = \cos\delta \, A(n^*,\delta) \, I(z,\beta) \tag{5-38}$$

where $A(z,\beta)P_{Laser}$ is the absorbed beam intensity, β is the circular angle, z is the depth from the top surface, and $I(z,\beta)$ is the incident beam intensity. An overall absorptivity for the erosion front can be found by integrating the absorbed laser power over the erosion front area and dividing by the total laser power. The erosion front shape can be determined through a power balance between laser power and the rates of heating, melting, and thermal loss. Material removal is assumed to occur through melting completely; no material is vaporized.

$$A(z,\beta)P_{Laser} = P_T + P_m + P_l \tag{5-39}$$

where P_{Laser} is the total laser power and the heating power is

$$P_T = b_s S V \rho c (T_p - T_0) \tag{5-40}$$

the melting power is

$$P_m = b_s S V \rho \varepsilon_m \tag{5-41}$$

and the heat conduction loss is

$$P_l = \frac{\pi k (T_m - T_0) \sqrt{b_s s}}{\left(\arctan\dfrac{16\kappa}{Vs}\right)^{1/2}} \exp\left(-\frac{Vb_s}{2\kappa}\right) \tag{5-42}$$

In the power balance, s is the cutting depth, b_s is the kerf width, T_m is the melting temperature, T_0 is the ambient temperature, and V is the scanning velocity. Material properties include the thermal conductivity k, thermal diffusivity κ, density ρ, specific heat c, melting temperature T_m, and heat of fusion ε_m. The power balance in Eq. (5-42) can be solved for s to determine the shape of the erosion front.

A coaxial gas jet is usually used in tandem with the laser beam to protect the focussing lens and remove molten and plasma material during laser cutting. In [52] the forces exerted by the gas jet on the molten layer in laser cutting were investigated theoretically by solving the equations of motion of the gas flow. It was reported that momentum is transferred from the gas jet to the erosion front by a pressure gradient and friction, and both effects are of the same order. It was assumed that the gas flow is laminar within the cutting kerf, and the flow is subsonic. This assumption is not proper, since for most gas jet operating pressures the flow is supersonic.

In order to investigate the appropriate relationship between cutting speed, laser power and workpiece thickness, a model is proposed based on the assumption that material removal occurs at a nearly vertical surface at the momentary surface of the cut (erosion front) [45]. The erosion front is covered by a liquid layer and joined by a molten layer (Fig. 5.4). This molten layer is heated by the absorbed laser radiation by reaction between the impinging gas particles and the molten material, and cooled by vaporization and by ejection of liquid material at the lower surface of the workpiece, by heat conduction, and by melting of solid material due to movement of the end of the cut in the direction of the cutting speed. This material removal takes place via ejection of molten material at the lower surface of workpiece and by evaporation from surface of melt. A mathematical analysis of reactive-gas laser cutting process requires a set of constitutive heat transfer and fluid mechanics relationships. These relationships include balances for the gas and material particles in the molten layer, the heat conduction equation, the energy, momentum and the mass balance of the liquid layer.

The mass balance for the reactive gas particles on the melt comprises of a gain in reactive gas particles through material melting countered by losses gas mass losses through chemical reactions (with reaction rate k_R), by ejection of liquid material at lower surface (with velocity v_s) and by evaporation described by the temperature dependent quantity α_R. Thus,

$$bdq_R = sbdk_R n_R n_A + sbn_R v_s + bd\alpha_R v_R \qquad (5\text{-}43)$$

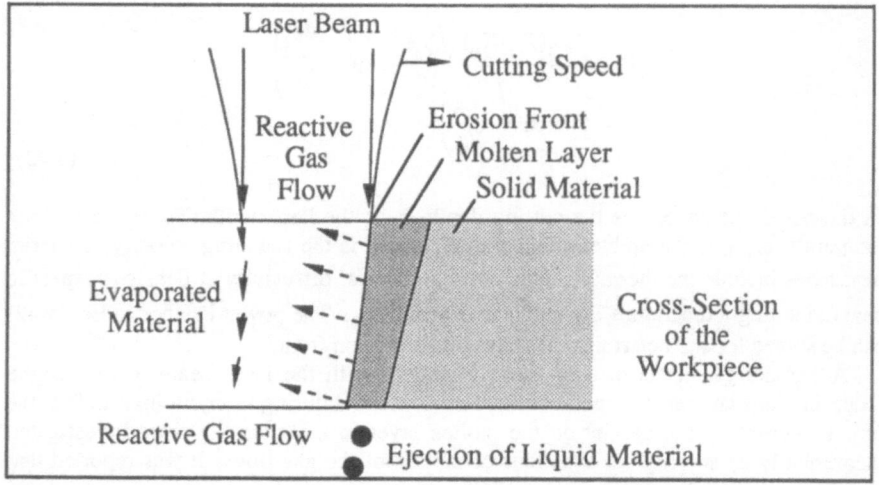

FIGURE 5.4   Physical Mechanism of Reactive Gas Assisted Laser Cutting

where s is the molten layer thickness, b is the kerf width, d is the workpiece thickness, n_A is the solid material density, n_R is the gas density, v_R is the reaction speed, q_R is the heat of reaction, and α_R is the rate of evaporation. The balance of pure metal particles in the melt comprises of gain due to melting of solid material (during the proceeding of the molten region) and similar to Eq. (5-43), losses by reaction with rate k_R, by ejection of liquid material at lower surface of cut (with velocity v_s) and by evaporation described by temperature dependent quantity α_A. Therefore,

$$bdvn_0 = sbdk_R n_R n_A + sbn_A v_s + bd\alpha_A n_A \qquad (5\text{-}44)$$

where n_0 is the density of the solid material. From Eq. (5-44), the number of reaction events per unit time in the molten layer and consequently, the heat produced by reaction in molten layer is calculated

$$\left(\frac{dE}{dt}\right)_R = \varepsilon_R bdvn_0 \frac{k_R n_R}{k_R n_R + \dfrac{v_s}{d} + \dfrac{\alpha_A}{s}} \qquad (5\text{-}45)$$

where ε_R is the average energy removed through evaporation. A large part of the reactive gas particles impinging on the surface of the melt is reflected or diffused back and carries on average an energy ε_{th} away from the melt. That energy loss is given by the fraction of impinging gas particles absorbed by surface of molten layer

$$\left(\frac{dE}{dt}\right)_{th} = \varepsilon_{th}(1 - \beta)q_0 = \varepsilon_{th}\left(\frac{1 - \beta}{\beta}\right)q_R \qquad (5\text{-}46)$$

where β is the molten layer reflectivity. Hence, the net energy gain of the molten layer due to reactive gas flow is obtained

$$\left(\frac{dE}{dt}\right)_{netR} = \frac{\varepsilon_R bdvn_0 q_R}{q_R + \dfrac{s}{k_R}\left(\dfrac{v_s}{d} + \dfrac{\alpha_A}{s}\right)\left(\dfrac{v_s}{d} + \dfrac{\alpha_R}{s}\right)} - \varepsilon_{th}\left(\frac{1 - \beta}{\beta}\right)q_R \qquad (5\text{-}47)$$

The energy gain by reactions between the gas and material can be maximized through Eq. (5-47). The maximum energy gain is given by the number of material atoms entering the melt per unit time and by the reaction energy

$$\left(\frac{dE}{dt}\right)_{Rmax} = \varepsilon_R bdvn_o \tag{5-48}$$

The cooling effect of the reactive gas flow increases as the strength of the gas flow rises. Thus the net energy gain shows a maximum for a critical strength of the gas flow. The momentum of the molten layer in a vertical direction is increased by friction with the gas flow, and is decreased by ejection of the melt from the lower surface of the workpiece. An expression for the velocity of the ejected molten material is

$$v_s = \sqrt{\frac{\eta_R d}{\delta_s sb} v_R} \tag{5-49}$$

where δ_s is the density of the melt, η_R is the dynamic viscosity of the reactive gas. It can be assumed that at the erosion front a rotational symmetry exists. Hence the energy lost from the erosion front into the workpiece by heat conduction is

$$\left(\frac{dE}{dt}\right)_K = 2\pi dKT \frac{\exp(-vb/4\kappa)}{K_o(vb/4\kappa)} \tag{5-50}$$

where K is the thermal conductivity and K_o is a Bessel function. The following balance can be used with Eq. (5-50) and with the assumption of optimum net energy gain by reaction

$$P_L(1 - e^{-\alpha_L d}) + \varepsilon_R sbdk_R n_R n_A = 2\pi dKT \frac{\exp(-vb/4\kappa)}{K_o(vb/4\kappa)} + \varepsilon_v \delta\beta(\frac{133.3}{\sqrt{2\pi kTm_s}}10^B T^C 10^{A/T}) \tag{5-51}$$

Combining Eq. (5-49) and (5-51) results in the equation:

$$\frac{bv}{4\kappa} = \frac{b}{4\kappa\sqrt{vb/4\kappa}} \sqrt{\frac{\eta_R}{2\delta_s}} \sqrt{\frac{v_R}{d}} \sqrt{\frac{K_o(vb/4\kappa)}{K_1(vb/4\kappa)}} \sqrt{1 - \frac{T_s}{T}} + \frac{1}{n_o}(\frac{133.3}{\sqrt{2\pi kTm_s}}10^B T^C 10^{A/T}) \tag{5-52}$$

Eqs. (5-51) and (5-52), which describe the equilibrium energy and mass balances, describe relationships between the temperature of the molten layer, the cutting speed and the thickness of the workpiece, kerf width and absorbed laser power. Intersections between mass balances for $s=0$ and s_{max} with the energy balance determine the minimum and maximum cutting speeds that can be obtained for given laser power and material thickness. The minimum speed corresponds to material removal solely by evaporation (i.e., sublimation cutting) while an increasing cutting speed corresponds to a decreasing contribution of evaporation.

In a unique case, where the temperature distribution in the molten layer and in the adjacent solid material remains stationary, the thickness s of the molten layer can be determined from the boundary condition that the temperature at the boundary between the liquid and solid region must be equal to the melting point. Consequently,

$$s = \frac{2\kappa}{v}(1 - \frac{T_s}{T})\frac{K_o(vb/4\kappa)}{K_1(vb/4\kappa)}$$

(5-53)

With Eqs. (5-51-5.53), the contributions to material removal by evaporation and by melting can be be calculated from the given cutting speed, temperature, and thickness of the molten layer.

For many materials, periodic striations are observed on the cutting surfaces which degrade the surface quality. The striations on the cutting surfaces are due to an unsteady motion of the molten layer. When a laser beam heats up material to a temperature where it will ignite in the presence of oxygen, the resulting combustion pushes molten material radially away from the laser spot. The combustion front comes to rest and the subsequent encroachment of the laser beam on the new erosion front repeats the oxidation initiation process. The resulting cut surface consists of regularly spaced striations. This phenomenon suggests a new technique for improving surface quality by utilizing a specific range of laser pulsing frequencies. In [48] a dynamic solution predicted a periodic oscillation superimposed on the steady state temperature of the melt. A dynamic model of melt ejection due to friction forces and the pressure gradient of the gas flow by a gas jet was presented in [53].

Roughness of the cut edge is caused by periodic striations that are typical for laser cutting. Striation patterns on laser cuts determine to a large extent cut quality. When studying the dynamic behavior (i.e., roughness of the cut) of the kerf width, one must begin with the temperature in the center of the erosion front. If the temperature in the center of the erosion front is suddenly increased to T_{max}, the melting front begins to extend and reaches after a certain time an increased width b (T_{max}), the temporal growth of the width b can be assumed to be ruled by the following equation

$$\frac{db}{dt} = \frac{b(T_{max}) - b}{\tau}$$

(5-54)

where τ is the period between successive striations. The time to heat a volume of b + db to the melting point T_m is (u_b - length of isothermal):

$$dt = \frac{db\, u_b\, d\, c_v\, T_m}{\left(\dfrac{dE}{dt}\right)_{add}}$$

(5-55)

where the denominator term represents additional heat flow due to an increase of temperature on the erosion front. Moreover, the maximum temperature generated by a moving circular heat source with a Gaussian intensity distribution can be expressed by:

$$T_{max} = -E_i\left(\frac{v^2 r_f^2}{16 x^2}\right)\left(\frac{1}{4\pi K d}\right)\left(\frac{dE}{dt}\right)$$

(5-56)

Mathematical manipulation of Eqs. (5-54) and (5-56) yields the following expression for the initial moment after the temperature rise

$$\frac{db}{dt} = \frac{4\pi K d}{-E_i\left(\dfrac{v^2 r_f^2}{16 x^2}\right)}\left(\frac{T_{max,new} - T_{max,old}}{u_b\, d\, c_v\, T_m}\right)$$

(5-57)

Comparison of Eqs. (5-54) and (5-57) yields

$$\tau = \frac{\partial b}{\partial T_{max}}\left(-E_i\left(\frac{v^2 r_f^2}{16 x^2}\right)\right)\left(\frac{u_b\, c_v\, T_m}{4\pi K}\right)$$

(5-58)

If the kerf width, which is equal to the width of the liquid body, depends on time due to temporal changes of the temperature, then the following expressions for energy and mass balances are obtained:

$$A_1 PL + E_r n_o dvb - P_{loss}(b,t) = kc_v d\frac{ds}{dt}bT + kc_v ds\frac{db}{dt}T + kc_v dsb\frac{dT}{dt}$$

(5-59)

$$m_{gain} - m_{loss}(T,s,b) = k\rho_m sd\frac{db}{dt} + k\rho_m db\frac{ds}{dt}$$

(5-60)

where A_1 is the erosion front area, P is the laser power, and L is the latent heat of fusion. For treatment of small perturbations

$$A_1PL_1 = (-E_r n_o dv + \frac{\partial P_{loss}}{\partial b})b_1 + (c_v ds_o T_o k)\frac{db_1}{dt} + c_v db_o T_o k\frac{ds_1}{dt} + \frac{\partial P_{loss}}{\partial T}T_1 + c_v ds_o b_o k\frac{dT_1}{dt}$$

(5-61)

$$\frac{\partial m_{gain}}{\partial b}b_1(t) - \frac{\partial m_{loss}}{\partial T}T_1(t) - \frac{\partial m_{loss}}{\partial s}s_1(t) - \frac{\partial m_{loss}}{\partial b}b_1(t) = \rho_m ks_o d\frac{db_1}{dt} + \rho_m kdb_o\frac{ds_1}{dt}$$

(5-62)

Eqs. (5-61) and (5-62) and the differential equation for b(t) can be reduced to three algebraic equations:

$$A_1PL_1 = k_3 b_1 + k_5 s_1 + k_1 T_1$$

(5-63)

$$k_2 b_1 = k_4 s_1 + k_6 T_1$$

(5-64)

$$b_1 k_o = T_1$$

(5-65)

Eqs. (5-63-5.65) yield a final expression for the amplitude of the temperature:

$$T_1 = A_1PL_1 \frac{1}{k_1 + \frac{k_3}{k_o} + \frac{k_5}{k_4}(\frac{k_2}{k_o} - k_6)}$$

(5-66)

The resulting complex amplitude of the kerf width is:

$$b_1 = \frac{1}{k_o} T_1$$

(5-67)

The value of b_1 ultimately determines the depth of the striations.

Due to the movement of the erosion front with cutting speed v, the oscillation of the kerf width with frequency ω and amplitude $|b_1|$ causes periodic distortions of the cut edges. The subsequent wavelength of these striations is:

$$\lambda = \frac{2\pi v}{\omega}$$

(5-68)

5.2.3   Grooving

In grooving, a laser beam does not cut through the entire workpiece. The grooving process exhibits complicated characteristics, such as three-dimensional heat transfer, two material phases, a moving boundary, a spatially distributed heat source, etc. (Fig. 5.5). Little work has been done on the grooving process. A numerical analysis of laser grooving was done by assuming immediate evaporation of the solid material due to the laser irradiation in [37]. The governing equation in this study was a groove formation equation, and the temperature distribution inside the medium was assumed. A model which separates the process into different regions was suggested in [13]. Experimental results on laser grooving were reported in [18]; grooves in metallic and ceramic materials were produced using a single laser beam. As an element of three-dimensional laser machining, laser grooving was investigated in [15].

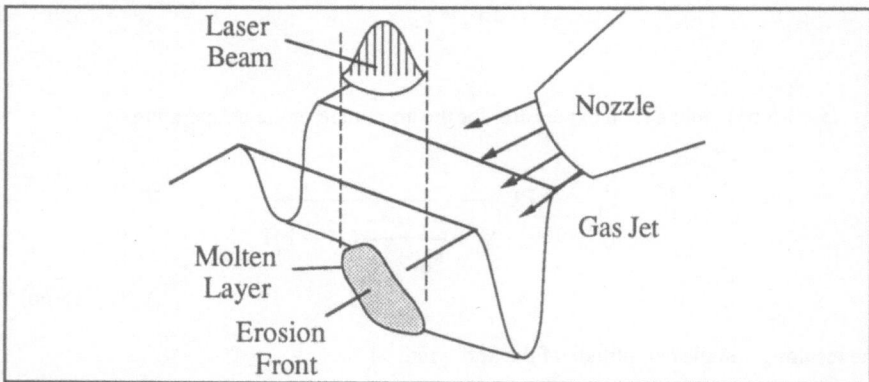

FIGURE 5.5   Schematic of Laser Grooving

A three-dimensional heat conduction equation governs the heat transfer problem with a moving heat source:

$$k\left(\frac{\partial^2 T}{\partial x^2} + \frac{\partial^2 T}{\partial y^2} + \frac{\partial^2 T}{\partial z^2}\right) = v\rho c_p \frac{\partial T}{\partial x}$$

(5-69)

or

$$\nabla^2 T = \frac{v}{\alpha}\frac{\partial T}{\partial x}$$

(5-70)

subject to:

$$x=\pm\infty, \; y=\pm\infty, \; z=\infty \; : \; T=T_0 \qquad (5\text{-}71)$$

where k is the conductivity, T is the temperature, v is the scanning velocity, ρ is the density, c_p is the specific heat, and α is the thermal diffusivity.

Modest et al. [37] studied the grooving process numerically by assuming evaporation due to laser irradiation on a moving semi-infinite solid. By considering the energy balance on a control surface which is *on* or *away* from the erosion front (Fig. 5.6), the following energy balance is derived:

$$aJ_0(\bar{n}\cdot\bar{k}) \, e^{-(x^2+y^2)/R^2} = h(T-T_0) - k(\bar{n}\cdot\nabla T) - \rho v L(\bar{i}\cdot\bar{n})$$

(5-72)

where a is the absorptivity, J_0 is the peak intensity, R is the beam radius, L is the latent heat of fusion, h is the convection heat transfer coefficient, and i, j, and k are the unit directional vectors for the cartesian coordinate system.

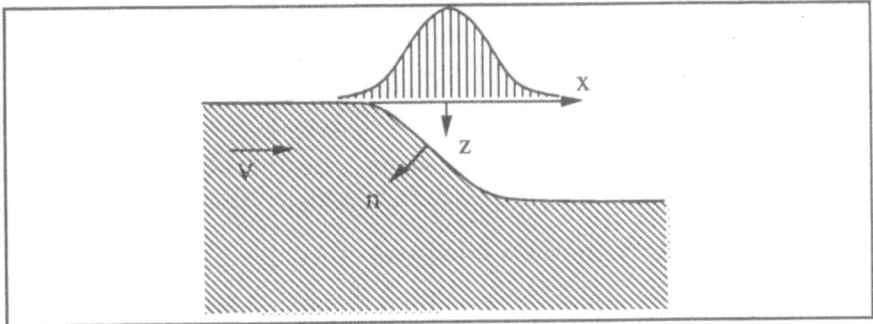

FIGURE 5.6   Control Surface on Erosion Front

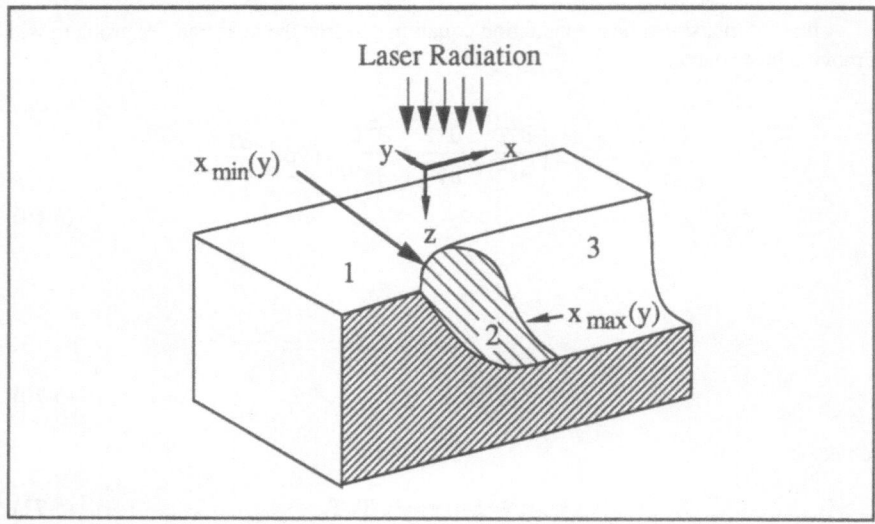

FIGURE 5.7   Different Regions in Laser Grooving

Equation (5-72) can be further modified by considering three distinct regions (Fig. 5.7):

- *Region 1* is the region which the laser beam has not yet passed. In this region, $x\ll0$, $i\bullet n=0$, $j\bullet n\neq0$, $n\neq k$, and $T<T_m$. s_- represents the depth of the surface in region 1.

$$aJ_o\ e^{-(x^2+y^2)/R^2} = (h(T-T_o)-k(\bar{n}\bullet\nabla T))\left(1+\left(\frac{\partial s_-}{\partial y}\right)^2\right)^{1/2}$$

(5-73)

- *Region 2* is the evaporation front where $i\bullet n\neq0$, $j\bullet n\neq0$, and $k\neq n$.

$$aJ_o e^{-(x^2+y^2)/R^2} = \left(h(T-T_o)-k(n\bullet\nabla T)\right)\left(1+\left(\frac{\partial s}{\partial x}\right)^2+\left(\frac{\partial s}{\partial y}\right)^2\right)^{1/2}+\frac{\rho vL}{2}\frac{\partial s}{\partial x}$$

(5-74)

Since the depth of the groove $s(x,y)$ is unknown in equation (5-74), this equation is the governing equation. The boundary conditions for equation (5-74) are

$$T=T_m,\ z=s(x,y)\ ,\ and\ x_{min}(y)<x<x_{max}(y)$$

(5-75)

- *Region 3* is the currently established groove with a depth of s(y). For this region i•n=0, j•n≠0, and k•n≠0. s_+ represents the surface of the fully-established groove.

$$aJ_o\, e^{-(x^2+y^2)/R^2} = (h(T - T_o) - k(\bar{n}\cdot\nabla T))\left(1 + \left(\frac{\partial s_+}{\partial y}\right)^2\right)^{1/2}$$

(5-76)

In order to arrive at solutions of the previously derived relationships, non-dimensional parameters can be introduced. Introducing non-dimensional parameters in the heat transfer and heat balance equations for each of the three regions, the following non-dimensional equations are derived, respectively:

$$U\frac{\partial\Theta}{\partial\xi} = \frac{\partial^2\Theta}{\partial\xi^2} + \frac{\partial^2\Theta}{\partial\eta^2} + \frac{\partial^2\Theta}{\partial\zeta^2} ; \quad -\infty < \xi,\ \eta < +\infty,\ S \leq \zeta < +\infty$$

(5-77)

subject to

$$\xi = \pm\infty,\ \eta = \pm\infty,\ \zeta = +\infty :\ \Theta = 0$$

(5-78)

where

$$\Theta = \frac{T - T_0}{T_m - T_0},\ \xi = \frac{x}{R},\ \eta = \frac{y}{R},\ \zeta = \frac{z}{R},\ \text{and}\ U = \frac{\rho c_p v R}{k}$$

(5-79)

Region 1

$$\zeta = S_-(\eta) :\ e^{-(\xi^2+\eta^2)} - N_k\left(Bi\Theta - \frac{\partial\Theta}{\partial\zeta}\right)\left(1 + \left(\frac{\partial S_-}{\partial\eta}\right)^2\right)^{1/2}$$

(5-80)

$$-\infty < \xi < \xi_{min}(\eta),\ -\infty < \eta < +\infty$$

(5-81)

where Bi is the Biot number (hx/k) and Nk = k(T_m-T_0)/(aJ_0R).

Region 2

$$\zeta = S(\xi,\eta) : \ N_e \frac{\partial S}{\partial \xi} = e^{-(\xi^2+\eta^2)} - N_k \left(Bi\Theta - \frac{\partial\Theta}{\partial n} \right) \left(1 + \left(\frac{\partial S}{\partial \xi}\right)^2 + \left(\frac{\partial S}{\partial \eta}\right)^2 \right)^{1/2}$$

(5-82)

$$\Theta = 1; \ \xi_{min}(\eta) < \xi < \xi_{max}(\eta), \ -\eta_{max} < \eta < \eta_{max}$$

(5-83)

where $N_e = h(T_m-T_0)/(aJ_0)$.

Region 3

$$\zeta = S_+(\eta) : \ e^{-(\xi^2+\eta^2)} - N_k \left(Bi\Theta - \frac{\partial\Theta}{\partial\zeta} \right) \left(1 + \left(\frac{\partial S_+}{\partial \eta}\right)^2 \right)^{1/2}$$

(5-84)

$$\xi_{min}(\eta) < \xi < +\infty, \ -\eta_{max} < \eta < \eta_{max}$$

(5-85)

By assuming, similar to the analytical approach, that heat is conducted primarily in the direction normal to the groove surface (i.e., $\nabla^2\Theta \approx \partial^2\Theta/\partial n^2$) and that a non-dimensional temperature profile can be approximated as $\Theta = \Theta_s(1-n/\delta)^2$, the above equations can be rewritten as:

Region 1:

$$\frac{\partial}{\partial \xi}(\Theta_s\delta) = \frac{6\Theta_s}{U\delta}$$

(5-86)

$$N_k\Theta_s \left(Bi + \frac{2}{\delta} \right) \left(1 + \left(\frac{\partial S}{\partial \eta}\right)^2 \right)^{1/2} = e^{-(\xi^2+\eta^2)}$$

(5-87)

Region 2:

$$\frac{\partial \delta}{\partial \xi} = \frac{6}{U\delta} - 3\frac{\partial S}{\partial \xi} \left(1 + \left(\frac{\partial S}{\partial \xi}\right)^2 + \left(\frac{\partial S}{\partial \eta}\right)^2 \right)^{-1/2}$$

(5-88)

$$N_e \frac{\partial S}{\partial \xi} = e^{-(\xi^2 + \eta^2)} - N_k \left(Bi + \frac{2}{\delta}\right)\left(1 + \left(\frac{\partial S}{\partial \xi}\right)^2 + \left(\frac{\partial S}{\partial \eta}\right)^2\right)^{1/2}$$

(5-89)

Region 3:

$$\frac{\partial}{\partial \xi}(\Theta_s \delta) = \frac{6\Theta_s}{U\delta}$$

(5-90)

$$N_k \Theta_s \left(Bi + \frac{2}{\delta}\right)\left(1 + \left(\frac{\partial S}{\partial \eta}\right)^2\right)^{1/2} = e^{-(\xi^2 + \eta^2)}$$

(5-91)

To determine the groove shape, the boundary between **Region 1** and **Region 2** has to be calculated. At the boundary $(T=T_m)$ material starts to melt. By defining a function Φ, for **Region 1**, as:

$$\Phi = N_k Bi\Theta_s e^{\eta^2}\left(1 + \left(\frac{\partial S}{\partial \eta}\right)^2\right)^{1/2}$$

(5-92)

and by combining Eqs. (5-86) and (5-87), one can write the following ordinary differential equation:

$$\frac{d\Phi}{d\xi} = \frac{\frac{3Bi^2}{2U}(e^{-\xi^2} - \Phi)^3 - 2\xi e^{-\xi^2}\Phi^2}{\Phi(2e^{-\xi^2} - \Phi)}$$

(5-93)

The function, Φ, can be found by integrating Eq. (5-93). The point $(\xi = \xi_{min}, \eta = 0)$ on the boundary can be determined.

$$\Theta_s = \frac{\Phi e^{-\eta^2}}{N_k Bi} = 1$$

(5-94)

Since

$$\left(1 + \left(\frac{\partial S}{\partial \eta}\right)^2\right)^{1/2} = 1 \text{ at } \xi = 0$$

(5-95)

the boundary can be determined as a function of ξ and $\eta(\xi)$ by solving equation (5-94) with the known function, Φ, $\Theta_s=1$, and given function S with respect to η.

In **Region 2**, the depth of the groove and the penetration depth at the nodes $\eta > \eta_j$ and $\xi < \xi_i$ are either known as boundary conditions or can be calculated.

The solution procedure is repeated at the interior nodes (inside the boundary) sweeping $\xi_{min}(\eta)$ to $\xi_{max}(\eta)$ for a certain decreasing step of η until the depth S no longer changes with respect to ξ. The function $\xi(\eta)$ is the boundary between **Regions 2** and **3**. **Region 3**, namely the established groove after the last pass of the laser beam, has a depth dependent only on h and can be calculated from the groove depth in **Region 2**.

An analytical approach to determining the relationship between the groove depth and process parameters is shown in [15]. Since the erosion front presents a complicated three-dimensional shape and is difficult to solve analytically, this approach assumes that the erosion front can be divided into a set of infinitesimally-small control surfaces (Fig. 5.8); each control surface is linear on each side with an inclination angle θ from the x axis and ϕ from the y axis. Energy is introduced into the control volume due to beam irradiation, and it is dissipated through material ablation and heat conduction, as shown by the heat balance relation:

$$\frac{aJ(dA)}{\sqrt{1 + \tan^2\theta + \tan^2\phi}} = -k\left(\frac{dT}{dn}\right)_{n=0} dA + \rho L v \tan\theta \, dxdy$$

(5-96)

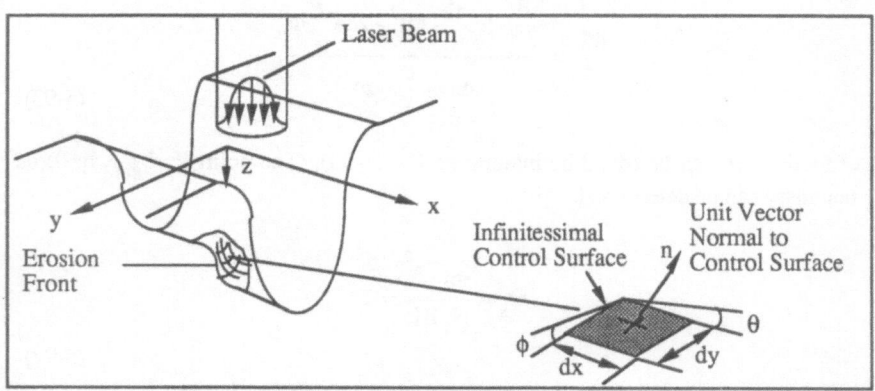

FIGURE 5.8   Control Surface for Analytical Model of Laser Grooving

where a is the absorptivity and L is the latent heat of fusion. Several assumptions are made for this analysis. First, beam-material interaction results in material removal at the vaporization temperature. Any molten material is assumed to be entirely removed by a gas jet and does not affect the beam-material interaction. Second, conduction into the workpiece occurs only normal to the groove surface. Third, the laser beam has a Gaussian intensity distribution:

$$J = \frac{P}{\pi R^2} \exp\left(- \frac{x^2 + y^2}{R^2} \right)$$

(5-97)

where P is the beam power. From the above assumptions, the integral relationship for groove depth can be derived from the control volume energy balance:

$$S(x,y) = \int_{-\infty}^{x} \frac{aP \exp\left(- \frac{x^2 + y^2}{R^2} \right)}{\pi R^2 \rho v (c_p (T_s - T_o) + L)} dx + S_+$$

(5-98)

For the case where multiple beam passes are used, S_+ is the surface created by the previous beam pass; for the case of only one beam pass S_+ is zero. The surface temperature T_s and the slope ϕ in the y direction are both functions of x and y. A general solution for the groove shape requires that ϕ and T_s be determined. However, the centerline change in groove depth $\Delta D = S(x,0) - S_+$ (where y=0 and ϕ=0) can be calculated without resorting to numerical methods. The incremental depth increase is:

$$\Delta D = \frac{2aP}{\pi^{1/2} \rho v d (c_p (T_s - T_o) + L)} = \frac{1.128aP}{\rho v d (c_p (T_s - T_o) + L)}$$

(5-99)

Based on Eq.(5-99), the total groove depth D can be calculated as $\lambda \Delta D$, where λ is the number of beam passes, assuming that the effect of beam divergence is negligible. This predicted groove depth can be expressed in log-log form as:

$$\log(D/d) = \log(P\lambda/vd) + \log(2a/\pi^{1/2}(\rho(c_p(T_s - T_o + L))))$$ (5-100)

The first right-hand-side term represents the groove depth dependence on the process parameters, while the second right-hand-side term represents the constant energy required to heat the material from the ambient temperature to the point of material removal.

5.2.4   Three-Dimensional Machining

Currently, conventional laser machining methods such as drilling, cutting and scribing are hindered by low energy efficiency and flexibility in terms of part geometry. In order to address these disadvantages, a laser machining concept has been developed [13] in which two laser beams create grooves in a workpiece. When the two grooves converge, a volume of material is removed (Fig. 5.9). According to the concept, the three-dimensional machining process increases flexibility by extending laser applications to three-dimensional shaping of parts. This concept is also substantially more energy efficient than single-beam ablation of the entire volume of material, since energy is only consumed for the grooves to be made and not for the entire volume of material to be removed.

The three-dimensional process can be decomposed into two laser grooving processes with interactions. Each groove is governed by a heat conduction equation similar to the case for single-beam grooving.

$$k\left(\frac{\partial^2 T}{\partial x^2} + \frac{\partial^2 T}{\partial y^2} + \frac{\partial^2 T}{\partial z^2}\right) = v\rho c_p \frac{\partial T}{\partial x} \qquad (5\text{-}101)$$

Since the two laser beams have similar spatial intensity distributions, the groove shapes created by the two laser beams will be similar and a plane of symmetry can be formed between the two grooves. Using the symmetry plane to isolate one groove, the domain for numerical analysis can be selected as shown in Figure 5.10. The boundary conditions are

$$\text{for} \quad x = x_{max}: \qquad\qquad \partial T/\partial x = 0 \qquad\qquad (5\text{-}102a)$$

$$\text{for} \quad x = x_{min}; y = y_{min}; y = y_{max}: \quad T = T_o \qquad\qquad (5\text{-}102b)$$

$$\text{for} \quad z = z_{min}(y): \qquad\qquad \partial T/\partial y = \partial T/\partial z \qquad\qquad (5\text{-}102c)$$

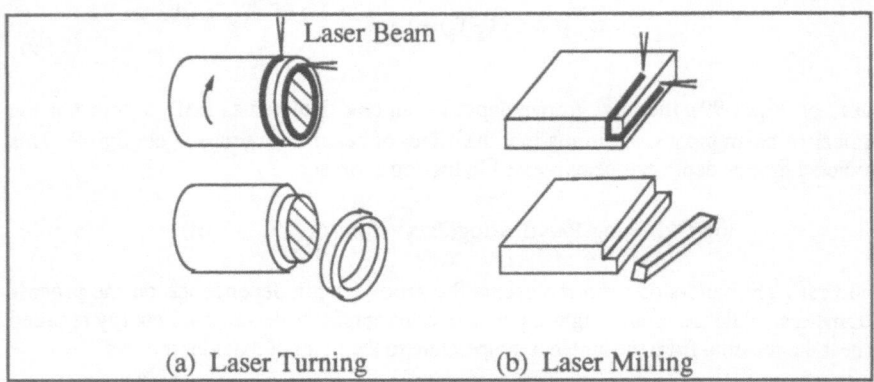

FIGURE 5.9   Three-Dimensional Laser Machining Concept

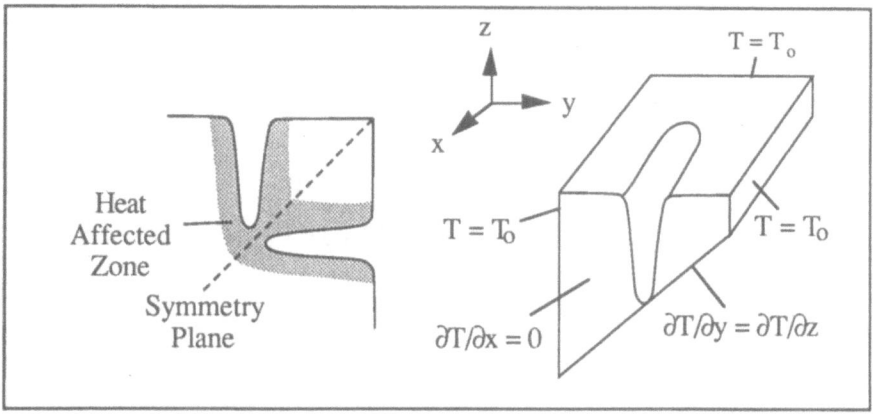

FIGURE 5.10   Numerical Domain and Boundaries for Three-Dimensional Machining

At the top surface ($z = z_{max}$), the energy balance at the erosion front is the boundary condition. In the control volume in Figure 5.10, the following energy balance can be derived for the erosion front

$$aJ(x,y,z)A_1 = - k \left(\frac{\partial T}{\partial n}\right) A + \rho L v A_2$$

(5-103)

where n is a coordinate normal to the surface pointing into the solid, A is the groove surface area, A_1 is the projected groove surface area in the x-y plane, and A_2 is the projected groove surface area on the plane of symmetry. From the geometry of the control volume, the following relations between the infinitesimal areas can be obtained:

$$\frac{A_1}{A} = \left(1 + \left(\frac{\partial s}{\partial x}\right)^2 + \left(\frac{\partial s}{\partial y}\right)^2\right)^{-1/2}$$

(5-104)

$$\frac{A_2}{A} = \left(\frac{\partial s}{\partial x}\right)\left(1 + \left(\frac{\partial s}{\partial x}\right)^2 + \left(\frac{\partial s}{\partial y}\right)^2\right)^{-1/2}$$

(5-105)

Substitution of Eqs. (5-104) and (5-105) into Eq.(5-103) yields:

$$\frac{a}{\gamma} J(x,y,z) = - k\left(\frac{\partial T}{\partial n}\right) + \frac{\rho L v}{\gamma}\left(\frac{\partial s}{\partial x}\right)$$

$$(5\text{-}106)$$

where:

$$\gamma = \left(1 + \left(\frac{\partial s}{\partial x}\right)^2 + \left(\frac{\partial s}{\partial y}\right)^2\right)^{1/2}$$

$$(5\text{-}107)$$

The numerical analysis accounts for the effect of beam defocussing, where the beam intensity decreases and the beam radius increases as the distance between the erosion front and the beam focal point increases. The Gaussian laser beam intensity, including the beam defocussing effect, can be expressed as:

$$J(x,y,z) = \frac{aP}{\pi r(z)^2} \exp\left(- \frac{x^2 + y^2}{r(z)^2}\right)$$

$$(5\text{-}108)$$

where assuming the laser beam is focused on the workpiece surface (z=0), the beam radius can be expressed as:

$$r(z) = R \left(1 + \left(\frac{\lambda_L z}{\pi R^2}\right)^2\right)^{1/2}$$

$$(5\text{-}109)$$

where λ_L is the beam divergence. Since the position of the erosion front has not been determined, the problem contains an unknown boundary condition. In this case the method of lines provides a convenient solution procedure; it has been developed for multi-dimensional heat transfer problems by Meyer [17]. According to the method of lines, Eq. (5-100) is transformed into a set of ordinary differential equations by replacing all derivatives with finite difference forms except for the derivative in the z-direction:

$$\frac{v}{\alpha}\left(\frac{T_{i+1,j} - T_{i-1,j}}{2\Delta x}\right) = \frac{T_{i+1,j} - 2T_{i,j} + T_{i-1,j}}{(\Delta x)^2} + \frac{T_{i,j+1} - 2T_{i,j} + T_{i,j-1}}{(\Delta y)^2} + \frac{\partial^2 T}{\partial z^2}$$

$$(5\text{-}110)$$

Eq. (5-110) is a boundary-value differential equation, requiring boundary conditions at z = z_{min} and the top surface (erosion front). Eq. (5-110) can be converted into initial-value differential equations by using the Riccati transformation. By defining a new function F,

$$\frac{\partial T}{\partial z} = \frac{1}{k} F(z)$$

(5-111)

Eq. (5-110) becomes:

$$\frac{\partial F}{\partial z} = \frac{k(T_{i+1,j} - 2T_{i,j} + T_{i-1,j})}{(\Delta x)^2} + \frac{k(T_{i,j+1} - 2T_{i,j} + T_{i,j-1})}{(\Delta y)^2} + \frac{\rho c v}{2\Delta x}(T_{i+1,j} - T_{i-1,j})$$

(5-112)

The Riccati transformation takes advantage of a relation between the functions F and T by introducing two other functions G(z) and H(z):

$$F(z) = G(z)T + H(z)$$

(5-113)

Application of the Riccati transformation results in three equations: for a forward sweep two equations start at $z = z_{min}$, and the position of the erosion front is determined; in a backward sweep, the temperatures within the domain are calculated. The two equations for the forward sweep are:

$$\frac{dG}{dz} = \frac{4k}{\Delta x^2} - \frac{1}{k} G^2$$

(5-114)

$$\frac{dH}{dz} = \frac{-k(T_{i+1,j} + T_{i-1,j})}{(\Delta x)^2} + \frac{k(T_{i,j+1} + T_{i,j-1})}{(\Delta y)^2} + \frac{\rho c v}{2\Delta x}(T_{i+1,j} - T_{i-1,j}) - \frac{GH}{k}$$

(5-115)

The initial values for G and H at the bottom plane can be determined from the boundary condition Eq. (5-103):

$$\frac{\partial T}{\partial z} = \frac{1}{k}(GT + H)$$

(5-116)

$$\frac{\partial T}{\partial y} = \frac{T - T_{-1}}{\Delta y}$$

(5-117)

$$GT + H = \frac{k}{\Delta y}(T - T_{-1})$$

(5-118)

where T_{-1} is the temperature at the node next to the node on the symmetric boundary in the y direction. Since Eq. (5-118) holds for arbitrary T, the initial values for G and H can be determined as:

$$G(z_{min}) = \frac{k}{\Delta y}$$

(5-119)

$$H(z_{min}) = - \frac{kT_{-1}}{\Delta y}$$

(5-120)

The temperature at the node which has the same y coordinate on the symmetric boundary as the node for the temperature T_{-1} is used for T_{-1}. In order to determine the groove depth for the given laser power and scanning velocity, the laser center is started from a small y value, and moved in the positive y direction until the groove bottom surface meets the symmetry boundary plane.

Since G in Eq. (5-114) is independent of temperature, the solution for G can be obtained prior to the calculation of the temperature and the surface profile. From the initial condition (5-119), G can be determined as:

$$G = A \left(\frac{1 + \left(\dfrac{G(z_{min}) - A}{G(z_{min}) + A} \right) e^{-2A(z-z_{min})/k}}{1 - \left(\dfrac{G(z_{min}) - A}{G(z_{min}) + A} \right) e^{-2A(z-z_{min})k}} \right)$$

(5-121)

where:

$$A = \left(2k^2 \left(\frac{1}{\Delta x^2} + \frac{1}{\Delta y^2} \right) \right)^{1/2}$$

(5-122)

The function H can be solved by integrating Eq. (5-115) from z_{min} to the top surface.

The energy balance equation at the top surface can be used as a criterion determining if ablation takes place. A function, C, is defined by arranging Eq. (5-106).

$$C = aJ - \rho Lv \left(\frac{\partial s}{\partial x} \right) - (GT_m + H) \left(1 + \left(\frac{\partial s}{\partial x} \right)^2 + \left(\frac{\partial s}{\partial y} \right)^2 \right)$$

(5-123)

At each node, C is calculated in the forward sweep to check if the energy balance (5-106) is satisfied. Since the energy balance (5-106) is satisfied at the erosion front, the erosion front is located where C vanishes. If C does not change sign before the previously made surface is reached, no ablation takes place. In this case, the surface temperature instead of the surface location is determined from the energy balance as follows:

$$\frac{a}{\beta^2} J = k(GT_s + H)$$

$$(5\text{-}124)$$

The equation for the backward sweep is:

$$\frac{\partial T}{\partial z} = \frac{1}{k} (GT + H)$$

$$(5\text{-}125)$$

The backward sweep starts at the top surface. At the erosion front of the top surface, the temperature is the melting temperature, T_m, and at the other surface the temperature is calculated from Eq. (5-116).

Iterations are repeated until the temperature and surface profile reach steady-state. During the calculation, updated temperatures, if available, are used. However, for the surface profile, the values calculated in the previous iteration are used, since a large slope can cause a numerical instability. A new temperature is obtained by an over-relaxation method:

$$T^m = T^{m-1} + \omega(T - T^{m-1})$$

$$(5\text{-}126)$$

where m is the iteration number, and w is the relaxation factor.

5.3 A Generalized Model For Laser Machining

In order to be able to give upper bounds for depths of cut in various laser machining processes, a general model for the laser machining processes with material removal can be developed. The general model is developed based on a heat balance at a erosion front and a temperature calculation inside a material from heat conduction equation.

Figure 5.11 shows the three processes and heat balances at the erosion front surfaces. The heat balance of the drilling process contains a non-steady term, $\partial s/\partial t$, since local material removal depends on the laser beam intensity and conduction heat.

In order to control the processes, simple relations between the depth of cut and process variables are needed for process models. The framework for a general model for the three processes is presented. The three processes involve three-dimensional heat transfer characteristics. In order to derive simple analytical relations, simplifications need to be made. For instance, in laser drilling the drilling direction is dominant over other directions. Due to the presence of a bottom surface in cutting which behaves as an adiabatic boundary, the heat conduction occurs two-dimensionally (downward conduction is negligible compared with conductions in other directions). Thus, drilling and cutting are treated as one- and two-dimensional processes, respectively.

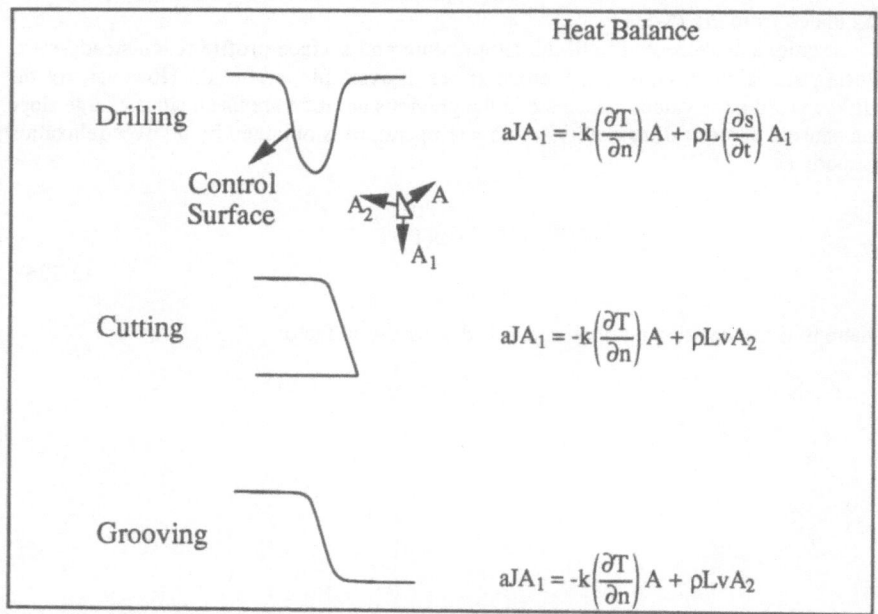

FIGURE 5.11   Heat Balances at Control Surfaces for Three Processes

When a molten layer has a negligible thickness, the absorbed laser beam power is used either to melt material or to be conducted into the solid.

5.3.1 Drilling

Drilling can be divided into two stages: heating and drilling stages (Fig. 5.12). In the heating stage, the temperature of the workpiece surface is increased up to the phase transition temperature by a laser beam interaction. The heating stage is usually very short because the laser beam intensity is very high. In the drilling stage, the hole depth is increased through molten material removal.

During the heating stage, the workpiece surface is not thermally eroded. It is difficult to obtain a simple analytical solution for drilling as a non-steady process with three-dimensional heat transfer characteristics. Thus, it is assumed that drilling is a one-dimensional process and that the laser beam intensity is uniform. The boundary conditions are

at $z = 0$

$$- k \left(\frac{dT}{dz} \right)_{z=0} = J_o$$

(5-127)

$$\text{as } z \rightarrow \infty \quad T = T_o$$

(5-128)

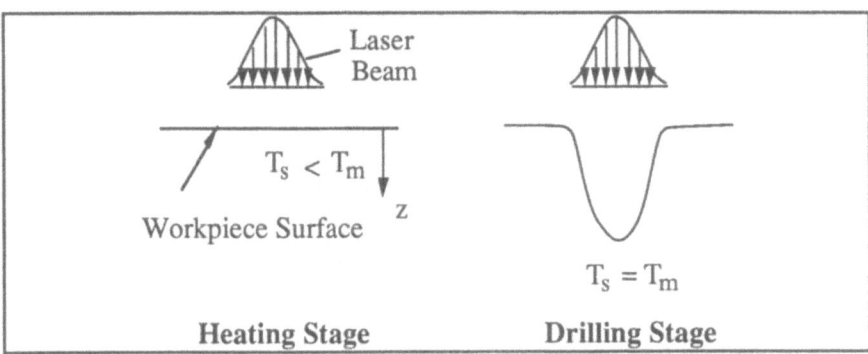

FIGURE 5.12   Heating and Drilling Stages

Given these two boundary conditions, the temperature distribution inside the workpiece can be derived from Eq. (5.4) as:

$$T - T_o = \frac{2J_o}{k}\left(\frac{\alpha t}{\pi}\right)^{1/2} e^{-z^2/4\alpha t} - \frac{J_o z}{k}\left(1 - \text{erf}\frac{z}{2(\alpha t)^{1/2}}\right)$$

(5-129)

where α is the thermal diffusivity and J_0 is the beam intensity. This temperature distribution is valid under the condition $(\alpha t)^{1/2} < R$, which can be achieved either through low diffusivities or small drilling times. The time for the workpiece surface to reach the phase transition temperature T_s can be determined from Eq. (5-129). Applying $T = T_s$ at $z = 0$, the following relation can be obtained.

$$T_s - T_o = \frac{2J_o}{k}\left(\frac{\alpha t}{\pi}\right)^{1/2}$$

(5-130)

The duration of the heating stage can be calculated as

$$t_h = \frac{\pi}{\alpha}\left(\frac{k(T_s - T_o)}{2J_o}\right)^2$$

(5-131)

During the heating stage, a hole is not made because phase transition does not occur. After the surface temperature reaches the melting point, drilling starts.

In order to determine the hole depth as a function of time and process variables, a one-dimensional analysis is considered. Figure 5.13 shows one-dimensional drilling. A laser beam is assumed to have a uniform intensity distribution J_o $(= P/\pi d^2)$.

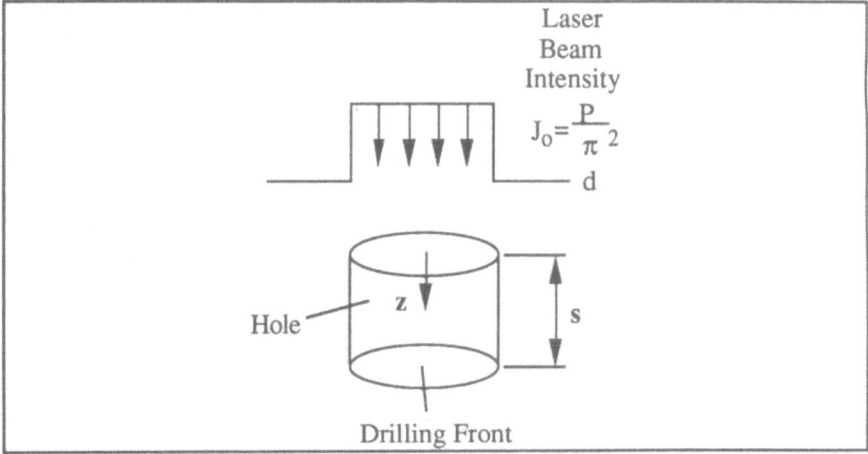

FIGURE 5.13   One-Dimensional Drilling

The heat balance at the erosion front can be expressed as

$$aJ_o = \rho L \frac{\partial s}{\partial t} - k \left(\frac{dT}{dz}\right)_{z=0}$$

(5-132)

where a is the absorptivity of the material, ρ is the density, L is the latent heat of fusion, k is the thermal conductivity, and T is temperature. In order to determine the drilling velocity ($\partial s/\partial t$), the temperature gradient at the drilling front should be known. The temperature distribution inside the solid is governed by the following heat conduction equation.

$$\frac{1}{\alpha} \frac{\partial T}{\partial t} = \frac{\partial^2 T}{\partial z^2}$$

(5-133)

where α is the thermal diffusivity. The heat conduction equation can be simplified as

$$-\frac{1}{\alpha} \left(\frac{\partial s}{\partial t}\right) \frac{dT}{dz} = \frac{d^2 T}{dz^2}$$

(5-134)

The boundary conditions are

$$T = T_s \text{ at } z = 0 \qquad (5\text{-}135)$$
$$T = T_0 \text{ at } z \to \infty \qquad (5\text{-}136)$$

By applying the boundary conditions, Eq. (5-134) can be solved for the temperature distribution inside the solid.

$$\frac{T - T_o}{T_s - T_o} = e^{-\frac{1}{\alpha}\left(\frac{ds}{dt}\right) z}$$

$$(5\text{-}137)$$

The temperature gradient at the drilling front can be determined as

$$\left(\frac{dT}{dz}\right)_{z=0} = -\frac{1}{\alpha}\left(\frac{ds}{dt}\right)(T_s - T_o)$$

$$(5\text{-}138)$$

Substitution of the temperature gradient into the energy balance yields

$$aJ_o = \rho L \left(\frac{ds}{dt}\right) + \rho c_p \left(\frac{ds}{dt}\right)(T_s - T_o)$$

$$(5\text{-}139)$$

The drilling velocity can be expressed as

$$\frac{ds}{dt} = \frac{aJ_o}{\rho\,(L + c_p(T_s - T_o))}$$

$$(5\text{-}140)$$

The hole depth can be determined by integrating Eq. (5-140).

$$t \le t_h \quad s = 0 \qquad (5\text{-}141)$$

$$t > t_h$$

$$s = \frac{aJ_o(t - t_h)}{\rho\,(L + c_p(T_s - T_o))} = \frac{4aP\,(t - t_h)}{\pi\rho d^2(L + c_p(T_s - T_o))}$$

$$(5\text{-}142)$$

Eq. (4-179) shows that the hole depth is proportional to the laser beam power and the beam interaction time.

5.3.2 Cutting

In order to gain a quantitative understanding of the effect of the different process parameters on the cutting process, an infinitesimal control surface on the erosion front surface can be studied (Figure 5-14). The control surface is inclined at an angle θ with respect to the x-axis and at an angle ϕ with respect to the y-axis, and is subjected to a laser beam of intensity J(x,y). The Cartesian coordinate system (x, y, z) is moving with the laser beam which has an intensity profile J(x,y) projected onto the groove surface. The heat balance at the control surface shown in Figure 5.14 is

$$aJ_o\, e^{-(x^2+y^2)/R^2}\,dxdy = \rho Lv\, dxdy\, \tan\theta - k\left(\frac{dT}{dn}\right)_{n=0} dxdy\, (1 + \tan^2\theta + \tan^2\phi)^{1/2}$$

$$(5\text{-}143)$$

where v is the scanning velocity, and n is a coordinate normal to the cutting surface.

Although heat is conducted three-dimensionally near the erosion front, due to the workpiece thickness (bottom surface is an adiabatic boundary) heat is conducted parallel to the bottom surface. Thus it is assumed that heat is conducted two-dimensionally into the solid. The conduction term in Eq. (5-143) can be simplified as

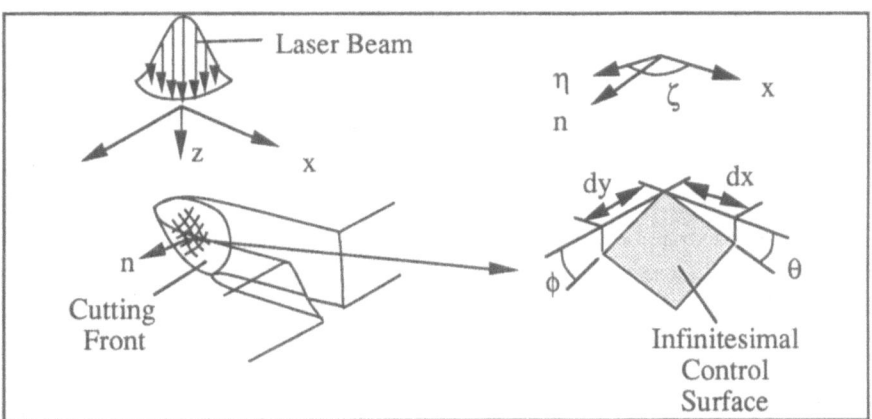

FIGURE 5.14   Control Surface on Cutting Front in Cutting Process

$$\left(\frac{dT}{dn}\right)_{n=0} (1 + \tan^2\theta + \tan^2\phi)^{1/2} = \left(\frac{dT}{d\eta}\right)_{\eta=0} \tan\theta$$

$$(5\text{-}144)$$

where η is a coordinate parallel to the bottom surface. Thus the heat balance (5-143) can be rewritten as

$$aJ_o\, e^{-(x^2 + y^2)/R^2} = \rho L v\, \tan\theta - k\left(\frac{dT}{d\eta}\right)_{h=0} \tan\theta$$

$$(5\text{-}145)$$

The temperature gradient at the erosion front can be determined by solving the following heat conduction equation.

$$\frac{v}{\alpha}\frac{\partial T}{\partial x} = \nabla^2 T$$

$$(5\text{-}146)$$

It is assumed that conduction area and direction do not change. According to this assumption, the terms in Eq. (5-146) can be simplified as

$$\frac{\partial T}{\partial x} = -\frac{\partial T}{\partial \eta}\cos\zeta$$

$$(5\text{-}147)$$

$$\nabla^2 T = \frac{\partial^2 T}{\partial \eta^2}$$

$$(5\text{-}148)$$

Eq. (5-146) can be transformed into a one-dimensional differential equation.

$$-\frac{v}{\alpha}\cos\zeta\,\frac{dT}{d\eta} = \frac{d^2T}{d\eta^2}$$

$$(5\text{-}149)$$

The boundary conditions for the cutting process are

$$\text{at } h = 0 \qquad T = T_s \qquad\qquad (5\text{-}150)$$
$$\text{as } h \to \infty \qquad T = T_0 \qquad\qquad (5\text{-}151)$$

The temperature distribution inside the solid can be determined as

$$\frac{T - T_o}{T_s - T_o} = e^{-\frac{v}{\alpha} \cos\zeta \, \eta}$$

(5-152)

Differentiation of the temperature distribution yields the following temperature gradient at the erosion front

$$\left(\frac{dT}{d\eta}\right)_{\eta=0} = -\frac{v}{\alpha} \cos\zeta \, (T_s - T_o)$$

(5-153)

By substituting the temperature gradient into the heat balance, Eq. (5-145) can be rewritten as

$$aJ_o \, e^{-(x^2 + y^2)/R^2} = \rho L v \tan\theta + \rho c_p \cos\zeta \, (T_s - T_o) \tan\theta$$

(5-154)

The maximum cutting depth can be achieved along the center line, where $y = 0$ and $\zeta = 0$. The erosion front slope in the cutting direction can be expressed as

$$\tan\theta = \frac{aJ_o \, e^{-x^2/R^2}}{\rho v (L + c_p(T_s - T_o))}$$

(5-155)

The infinitesimal depth is

$$ds = dx \tan\theta \qquad\qquad (5\text{-}156)$$

The cutting depth can be determined as a function of x by integrating from $-\infty$ to $+\infty$.

$$s = \int_{-\infty}^{\infty} ds = \int_{-\infty}^{\infty} \tan\theta \, dx$$

(5-157)

By substituting the expression for $\tan\theta$, Eq. (5-155), into Eq. (5-157), the following expression for the depth s can be obtained.

$$s = \int_{-\infty}^{\infty} \frac{aP\, e^{-x^2/R^2}}{pR^2(L + c_p(T_s - T_o))}\, dx$$

(5-158)

The temperature at the top surface T_s along the center line of the groove is assumed to be the melting temperature T_m. Although T_s varies from T_m at the erosion front to T_o far away from the erosion front, the resulting error in the groove depth should be negligible because the exponential term in Eq. (5-158) becomes negligible for the part of the surface where the temperature is not T_m. Consequently, s can be obtained as:

$$s = \frac{aP\pi^{1/2}R}{\pi R^2 \rho v(c_p(T_s - T_o) + L)} = \frac{2aP}{\pi^{1/2}\rho vd(c_p(T_s - T_o) + L)}$$

(5-159)

where d is the beam spot diameter, $d \equiv 2R$. The cutting depth is proportional to P/vd, which is the energy input per unit area of the workpiece. Also, the cutting depth is small for materials with a high melting point and a high latent heat of evaporation.

5.3.3   Grooving

Figure 5.15 schematically shows the process of laser grooving. Since the grooving process is similar to the cutting process except for workpiece thickness, a similar approach is used to derive the relation between groove depth and process variables. The heat balance at the control surface in Figure 5.15 is given by

$$aJ_o\, e^{-(x^2+y^2)/R^2}\, dxdy = \rho Lv\, dxdy\, \tan\theta - k\left(\frac{dT}{dn}\right)_{n=0} dxdy\, (1 + \tan^2\theta + \tan^2\phi)^{1/2}$$

(5-160)

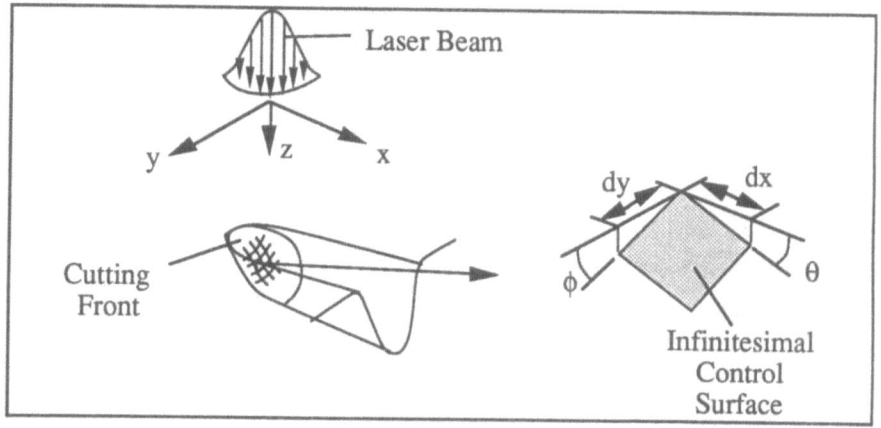

FIGURE 5.15   Analytical Model for the Laser Grooving Process

In order to determine the temperature gradient, the following heat conduction equation should be solved.

$$\nabla^2 T = \frac{v}{\alpha} \frac{\partial T}{\partial x}$$

(5-161)

subject to:

at the erosion front surface : $T = T_s$ (5-162)

for $x \to \pm\infty$, $y \to \pm\infty$, $z \to \infty$: $T = T_0$ (5-163)

In order to simplify Eq. (5-161), it is assumed that heat is conducted in the direction normal to the surface of the erosion front. Accordingly the following relations can be derived.

$$\frac{\partial T}{\partial x} = -\frac{\partial T}{\partial z} \tan\theta$$

(5-164)

$$\frac{\partial T}{\partial y} = \frac{\partial T}{\partial z} \tan\phi$$

(5-165)

$$\frac{\partial T}{\partial n} = \left(\left(\frac{\partial T}{\partial x} \right)^2 + \left(\frac{\partial T}{\partial y} \right)^2 + \left(\frac{\partial T}{\partial z} \right)^2 \right)^{1/2}$$

(5-166)

where n is a coordinate normal to the erosion front surface. Additionally, the following simplification is assumed.

$$\nabla^2 T = \frac{\partial^2 T}{\partial n^2}$$

(5-167)

From the above equations, $\partial T/\partial x$ can be related to $\partial T/\partial n$.

$$\frac{\partial T}{\partial x} = - \frac{\partial T}{\partial n} \frac{\tan\theta}{(1 + \tan^2\theta + \tan^2\phi)^{1/2}}$$

$$= - \beta \frac{\partial T}{\partial n}$$

(5-168)

$$\text{where} \quad \beta = \frac{\tan\theta}{(1 + \tan^2\theta + \tan^2\phi)^{1/2}}$$

(5-169)

Eq. (5-167) can be simplified as:

$$- \frac{\beta v}{\alpha} \frac{\partial T}{\partial n} = \frac{\partial^2 T}{\partial n^2}$$

(5-170)

subject to the following boundary conditions:

$$\text{at } n=0: \quad T=T_s \quad\quad (5\text{-}171)$$
$$\text{as } n\rightarrow\infty: T=T_o \quad\quad (5\text{-}172)$$

The temperature distribution inside the medium can be found as:

$$\frac{T - T_s}{T_s - T_o} = e^{-\frac{\beta v}{\alpha} n} - 1$$

(5-173)

The temperature gradient at the erosion front can be determined as:

$$\left(\frac{\partial T}{\partial n}\right)_{n=0} = -\frac{\beta v}{\alpha}(T_s - T_o)$$

(5-174)

Substitution of the temperature gradient into the heat balance yields

$$aJ = \frac{kv}{\alpha}(T_s - T_o)\tan\theta + \rho Lv\tan\theta$$

$$= \rho v\tan\theta(c_p(T_s - T_o) + L)$$

(5-175)

The slope of the groove can be determined as:

$$\tan\theta = \frac{\dfrac{aP}{\pi R^2}e^{-(x^2+y^2)R^2}}{\rho v(c_p(T_s - T_o) + L)}$$

(5-176)

Since the method for temperature estimation in grooving and cutting are the same, the slope for grooving is the same as that found in cutting. Thus the groove depth is the same as the maximum cutting depth for a given set of operating conditions.

$$s = \frac{aP\pi^{1/2}R}{\pi R^2 \rho v(c_p(T_s - T_o) + L)} = \frac{2aP}{\pi^{1/2}\rho vd(c_p(T_s - T_o) + L)}$$

(5-177)

References

1.   Atanasoc, P.A., and S.I. Gendjov, "Laser Cutting of Glass Tubing: A Theoretical Model," *J. Phys. D: Applied Physics*, Vol. 20 (1987) 597-601.

2.   Barber, R., "Hole Drilling with Lasers," *Creative Manufacturing Engineering Programs*, Rept. No. MR74-951.

3.   Belov, I.A., I.P. Ginzburg, and L.I. Shub, "Supersonic Underexpanded Jet Impingement upon Flat Plate," *International Journal of Heat Mass Transfer*, Vol. 16 (1973) 2067-2076.

4.   Brugger, K., "Exact Solutions for the Temperature Rise in a Laser-Heated Slab," *Journal of Applied Physics*, Vol. 43, No. 2 (Feb. 1972), 577-583.

5.   Bunting, K.A., and G. Cornfield, "Toward a General Theory of Cutting:A Relationship Between the Incident Power Density and the Cut Speed," *Journal of Heat Transfer* (Feb 1975), 116-122.

6.   Bush, A.J., and F.J. Kromer, "Simplification of the Hole-Drilling Method of Residual Stress Measurement," *ISA Transactions*, Vol. 12, No. 3 (Dec. 1973), 249-259.

7.   Chryssolouris, G., et al., "Theoretical Aspects of a Laser Machine Tool," *Journal of Engineering for Industry*, ASME, Vol. 110, No. 1 (Feb. 1988), 65-70.

8.   Chryssolouris, G., and J. Bredt, "Machining of Ceramics Using a Laser Lathe," *International Ceramic Review*, Vol. 37, No. 2 (1988), 43-45.

9.   Chryssolouris, G., and J. Bredt, and S. Kordas, "Laser Turning for Difficult to Machine Materials," *Proceedings of the Simposium on Machining of Ceramic Materials and Components*, Amer. Soc. of Mech. Eng., Vol. 17 (Nov. 1985), 9-17.

10.   Chryssolouris, G., and J. Bredt, and S. Kordas, "A New Machine Tool Concept Based on Lasers," *Proceedings of the XIV North Americal Manufacturing Research Conference/NAMRC XIV*, Soc. of Mfg. Eng. (May 1986), 245-250.

11.   Chryssolouris, G., and J. Bredt, "Laser Turning of Steels," *2nd Biennial International Machine Tool Research Forum* (Sept. 1987).

12.   Chryssolouris, G., and W.C. Choi, "Gas Jet Effects on Laser Cutting," *SPIE Conf. on High Power Lasers* (Jan. 1989).

13. Chryssolouris, G., and W.C. Choi, "Theoretical Aspects of Laser Grooving," *Proceedings, 14th Conference on Production Research and Technology* (Jan. 1987), 323-331.

14. Chryssolouris, G., W.C. Choi, S.B. Kyi, and P. Sheng, "Investigation of the Effects of a Gas Jet on Laser Grooving," *Proceedings of the XVI North Americal Manufacturing Research Conference/NAMRC XVI*, Soc. of Mfg. Eng. (May 1987), 217-222.

15. Chryssolouris, G, P.S. Sheng, and W.C. Choi, "Analysis on the Laser Machining Process for Ceramics and Composite Materials," *Proceedings, 15th Conference on Production Research and Technology* (Jan. 1989).

16. Cockayne, B., and D.B. Gasson, "The Machining of Oxides Using Gas Lasers," *Journal of Material Science*, Vol. 6 (1971), 126-129.

17. Comini, G., S.D. Guidice, R.W. Lewis, and O.C. Zienkiewics, "Fininte Element Solution of Non-Linear Heat Conduction Problems with Special Reference to Phase Change," *International Journal of Numerical Methods in Engineering*, Vol.8 (1974), 613-624.

18. Copley, S.M., M. Bass, and R.G. Wallace, "Shaping Silicon Compound Ceramics with a Continuous Wave Carbon Dioxide Laser," *Proceedings, Second International Symposium on Ceramic Machining and Finishing* (1978), 97-104.

19. Dabby, F.W., and U.-C. Paek, "High-Intensity Laser-Induced Vaporization and Explosion of Solid Material," *IEEE Journal of Quantum Electronics*, Vol. QE-8, No. 2 (Feb. 1972), 106-111.

20. Decker, I., J. Rue, and V. Atzert, "Physical Models and Technological Aspects of Laser Gas Cutting," *Proceedings of SPIE* (Sept. 1983), 81-88.

21. Duley, W.W., *Laser Processing and Analysis of Materials* , Plenum Press, New York, 1983.

22. El-Adawi, M.K., "Laser Melting of Solids-An Exact Solution for Time Intervals Less or Equal to the Transit Time," *Journal of Applied Physics*, Vol. 60, No. 7 (Oct. 1986), 2256-2265.

23. El-Adawi, M.K., and E.F. Elshehawey, "Heating a Slab Induced by a Time-Dependent Laser Irradiance-An Exact Solution," *Journal of Applied Physics*, Vol. 60, No. 7 (Oct. 1986), 2250-2255.

24. Eloy, J.-F., *Power Lasers* , Halsted Press, New York, 1987.

25. Fieret, J., and B.A. Ward, "Circular and Non-Circular Nozzle Exits for Supersonic Gas Jet Assist in CO_2 Laser Cutting," *Proceedings, Third International Conference on Lasers in Manufacturing (LIM3)*, (1986).

26. Gubanova, O.I., V.V. Lunev, and L.N. Plastinina, "The Central Breakaway Zone with Interaction between a Supersonic Unexpanded Jet and a Barrier," *Fluid Dynamics*, Vol. 6 (1973), 298-301.

27. Gummer, J.H., and B.L. Hunt, "The Impingement of Non-Uniform, Axisymmetric Supersonic Jets on a Perpendicular Flat Plate," *Israel J. Technology*, Vol. 12 (1974), 221-235.

28. Hamilton, D.C., and I.R. Pashby, "Hole Drilling Studies with a Variable Pulse Length CO_2 Laser," *Optics and Laser Technology* (Aug. 1979), 183-188.

29. Hassanein, A.M., and G.L. Kulcinski, "Simulation of Rapid Heating in Fusion Reactor Forst Walls Using the Green's Function Approach," *Journal of Heat Transfer*, Vol. 106 (Aug. 1984), 486-490.

30. Kobayashi, A., and Y. , "Laser Drilling of Nonmetals," *Toshiba Review* (Dec. 1971), 8-14.

31. Lee, C.S., A. Goel, and H. Osada, "Parametric Studies of Pulsed-Laser Cutting of Thin Metal Plates," *Journal of Applied Physics*, Vol. 58, No. 3 (Aug. 1985), 1339-1343.

32. Longfellow, J., "High Speed Drilling in Alumina Substrates with a CO_2 Laser," *Ceramic Bulletin*, Vol. 50, No. 3 (1971), 251-253.

33. Luxon, J., *Lasers in Manufacturing*, Prentice-Hall, Engelwood Cliffs, NJ, 1987.

34. Masters, J.I., "Problem of Intense Surface Heating of a Slab Accompanied by Change of Phase," *Journal of Applied Physics*, Vol. 27, No. 5 (May 1956), 477-484.

35. Miyazaki, T., "Drilling Characteristics of Metal Foil in Electron Beam Processing," *Bulletin of the Japan Society of Precision Engineering*, Vol. 13, No. 4 (Dec. 1979), 207-212.

36. Modest, M.F., and H. Abakians, "Heat Conduction in a Moving Semi-Infinite Solid Subjected to Pulsed Laser Irradiation," *Journal of Heat Transfer*, Vol. 108 (Aug. 1986), 597-607.

37. Modest, M.F., and H. Abakians, "Evaporative Cutting of a Semi-Infinite Body with a Moving CW Laser," *Journal of Heat Transfer* (Aug. 1986), 602-607.

38. Nakada, Y., and M.A. Giles, "X-Ray and Scanning Electron Microscope Studies of Laser-Drilled Holes in Al_2O_3 Substrates," *Journal of American Ceramic Society – Discussion and Notes,* Vol. 54, No. 7, 354-355.

39. Nielsen, S.E., *Laser Cutting with High Pressure Cutting Gases and Mixed Cutting Gases,* Ph.D. Thesis, Institute of Manufacturing Engineering, Technical University of Denmark (1985).

40. Pack, U.C., and F.P. Gagliano, "Thermal Analysis of Laser Drilling Processes," *IEEE Journal of Quantum Electronics,* Vol. QE-8, No. 2 (Feb. 1972), 112-119.

41. Petring, D., P. Abels, E. Beyer, and G. Herziger, "Werkstoffbearbeitung mit Laserstrahlung," *Feinwerktechnik & Messtechnik,* Vol. 96 (1988), 364-372.

42. Ready, J.F., "Effects Due to Absorption of Laser Radiation," *Journal of Applied Physics,* Vol. 36, No. 2 (Feb. 1965), 462-468.

43. Ruselowski, J.M., "Laser Selection for Cutting," *SME Technical Paper* (1987), Paper No. MR87-235.

44. Schulz, W., G. Simon, H.M. Urbassek, and I. Decker, "On Laser Fusion Cutting of Metals," *J. Phys. D: Applied Physics,* Vol. 20 (1987), 481-488.

45. Schuoecker, D., and W. Abel, "Material Removal Mechanism of Laser Cutting," *Proceedings of the SPIE* (Sept. 1983), 88-95.

46. Schuocker, D., and B. Walter, "Theoretical Model of Oxygen Assisted Laser Cutting," *Inst. Phys. Conf. Ser.,* No. 72 (Aug. 1984), 111-116.

47. Schuocker, D., "Theoretical Model of Reactive Gas Assisted Laser Cutting Including Dynamic Effects," *Proceedings of the SPIE,* Vol. 650 (1986), 210-219.

48. Schuocker, D., and P. Muller, "Dynamic Effects in Laser Cutting and Formation of Periodic Striations," *Proceedings of the SPIE,* Vol. 801 (1987), 258-264.

49. Schvan, P, and R.E. Thomas, "Time-Dependent Heat Flow Calculation of CW Laser-Induced Melting of Silicon," *Journal of Applied Physics,* Vol. 57, No. 10 (May 1985), 4738-4741.

50. Sparks, M., "Theory of Laser Heating of Solids: Metals," *Journal of Applied Physics,* Vol. 47, No. 3, (Mar. 1976), 837-849.

51. Stürmer, E., and M. von Allmen, "Influence of Laser-Supported Detonation Waves on Metal Drilling with Pulsed CO_2 Lasers," *Journal of Applied Physics,* Vol. 49, No. 11 (Nov. 1978), 5648-5654.

52.   Vicanek, M., and G. Simon, "Momentum and Heat Transfer of an Inert Gas Jet to the Melt in Laser Cutting," *J. Phys. D: Applied Physics.*, Vol. 20 (1987), 1191-1196.

53.   Vicanek, M., G. Simon, H. M. Urbassek, and I. Decker, "Hydrodynamical Instability of Melt Flow in Laser Cutting," *J. Phys. D: Applied Physics,* Vol. 20 (1987), 140-145.

54.   von Allmen, M., "Laser Drilling Velocity in Metals," *Journal of Applied Physics,* Vol. 47, No. 12 (Dec. 1976), 5460-5463.

55.   von Allmen, M., P. Blaser, K. Affolter, and E. Stürmer, "Absorption Phenomena in Metal Drilling with Nd-Lasers," *IEEE Journal of Quantum Electronics,* Vol. QE-14, No. 2 (Feb. 1978), 85-88.

56.   Waechter, D., P. Schvan, R.E. Thomas, and N.G. Tarr, "Modelling of Heat Flow in Multilayer CW Laser-Annealed Structures," *Journal of Applied Physics,* Vol. 59, No. 10 (May 1986), 3371-3374.

57.   Wagner, R.E., "Laser Drilling Mechanics," *Journal of Applied Physics,* Vol. 45, No. 10 (Oct. 1974), 4631-4637.

58.   Warren, R.E., and M. Sparks, "Laser Heating of a Slab Having Temperature-Dependent Surface Absorptance," *Journal of Applied Physics,* Vol. 50, No. 12 (Dec. 1979), 7952-7957.

6
Laser Machining Applications

This chapter presents laser machining results from industrial applications and laboratory experiments. In general, laser machining produces parts with higher dimensional accuracy and surface quality than those produced with conventional processes, and with higher material removal rates. Materials that can be machined by lasers include metals, ceramics, plastics, composites, wood, glass and rubber. *Laser drilling* can produce holes as small as 0.05mm in diameter in workpieces at a rate of up to 1ms/hole. It is used in industry for producing holes in turbine blades, combustion chambers, and aerosol nozzles, among other applications. *Laser cutting* is used to produce intricate two-dimensional shapes in workpieces made out of materials such as sheet metal and paper up to 15mm thick with high cutting speeds. *Laser scribing* has been used to create channels in ceramic substrates for cooling and identification labels in finished parts. Finally, *three-dimensional laser machining* has been used in a research effort to implement turning, milling and threading operations. Quality issues related to laser machining such as dross formation in metals, microcrack formation in ceramics and matrix decomposition in composites are also discussed in this chapter.

6.1 Introduction

The engineering materials most often used in industry can be divided into four categories: *metals*, *ceramics*, *polymers* and *composites*. *Wood*, *paper*, *rubber* and *glass* are also used in manufacturing. The selection of which manufacturing process to apply to a particular material is influenced by a number of factors:

- The *part geometry* will dictate the initial consideration of a manufacturing process; manufacturing processes can be characterized by their dimensional degrees of freedom such as one-dimensional drilling, two-dimensional cutting or three-dimensional turning. The complexity of a geometric feature will limit the number of processes capable of producing the final part shape.

- The *production volume* for a particular part determines both the flexibility of the equipment required and the production rate required. For machining processes, flexibility is determined by the variety of workpiece geometries the machine can handle, such as the number of machine axes and the range of part sizes the machine can accommodate. Production rate is a function of the material removal rate (MRR) of the process.

- The *physical properties* of the material, such as hardness, tensile strength, shear strength, melting/vaporization temperature, latent heat of fusion/vaporization, and thermal conductivity will place limitations on the type of process utilized.

- *Part quality* is usually expressed in terms of surface quality and dimensional accuracy. Surface quality is related to surface roughness, charring in the case of plastics or composites, micro-crack formation and heat-affected zone. Dimensional accuracy is related to the tolerances of the finished part in terms of straightness, degree of taper, accuracy of length and accuracy of angle.

In many cases, lasers offer a way to perform drilling, cutting or shaping operations where conventional machining processes fail. In other cases, laser machining offers an alternative material removal method which improves machining speed, surface quality and/or dimensional accuracy compared to traditional processes. Numerous laboratory studies and industrial applications of laser machining have taken place since the early 1960's (Fig. 6.1). This chapter summarizes the most significant experimental and industrial experience in laser machining (Table 6.1), including micromachining, namely machining with material removal rates lower tham $1mm^3$/min.

	Metals	Ceramics	Plastics	Composites	Others: Wood, Paper, Glass and Rubber
1-D Process/ Drilling	Section 6.2.1	Section 6.3.1	Section 6.4.1	Section 6.5.1	Section 6.6
2-D Process/ Cutting and Scribing	Sections 6.2.2, 6.2.3	Sections 6.3.2, 6.3.3	Sections 6.4.2, 6.4.3	Sections 6.5.2, 6.5.3	Section 6.6
3-D Process/ Turning, Milling and Threading	Section 6.2.4	Section 6.3.4	Section 6.4.4	Section 6.5.4	____

TABLE 6.1   Organization of Chapter 6

	Metals	Ceramics	Plastics	Composites	Others: Wood, Paper, Glass and Rubber
1-D Process/ Drilling	17, 32, 47, 50, 60	6, 14, 24, 29, 40, 56	14	14, 33	14, 81
2-D Process/ Cutting and Scribing	3, 4, 19, 23, 25, 30, 34, 39, 42, 44	1, 6, 10, 15, 16, 21, 24, 41, 45, 46	14, 41, 52, 53, 54, 58	11, 12, 22, 28, 48, 49, 51	2, 14, 26, 38, 59
3-D Process/ Turning, Milling and Threading	7, 35	1, 6, 15, 16	7	13, 51	—

FIGURE 6.1 Literature Overview for Laser Machining Applications (The numbers correspond to the references at the end of this chapter)

6.2 Metals

6.2.1 One-Dimensional Machining: Drilling

Lasers are widely used in metal drilling applications due to the high processing speeds, high tolerance, repeatability and minute dimensions achievable (Table 6.2). In most industrial applications, both the penetration speed of the laser beam and the positioning speed from hole to hole are important; therefore, the values for drilling rate are given as holes per second and represent the total cycle time per hole. Usually, the holes created have eccentricity values below 5% and surface roughness below 5μm [50]. Control of the drilling process for a given workpiece material is accomplished through selection of beam power, focal spot size, and drilling time. Laser drilling can be performed through continuous drilling, with the use of a continuous-wave beam, or percussion drilling, with the use of a pulsed beam. In continuous drilling, material removal occurs through melting with some vaporization (Fig. 6.2). The molten material is ejected out of the bottom of the hole with the aid of a gas jet. In percussion drilling, a pulsed beam removes material through melting and localized detonation or explosion. In this case, about 90% of the material is removed through detonation effects [32]. Additionally, a reactive gas jet (O_2) can be used to remove material through oxidation, chemical reactions, etc. [60, 61].

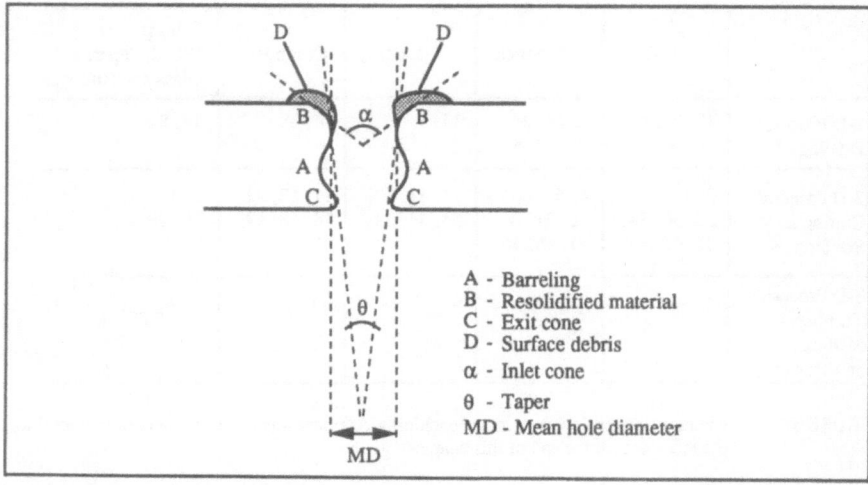

A - Barreling
B - Resolidified material
C - Exit cone
D - Surface debris
α - Inlet cone
θ - Taper
MD - Mean hole diameter

FIGURE 6.2   Features of Laser-Drilled Holes [61]

The effects of changing process variables such as gas jet pressure and lens/workpiece distance in laser drilling are studied in [60] using a pulsed Nd:Glass laser for stainless steel, titanium and nickel. Changes in gas pressure are brought about by the use of a vacuum on the hole exit side. Changes in lens/workpiece distance affect plasma formation inside the hole, which in turn affects the geometry and size of the melting front. Plasma formation is also affected by changes in pressure for the assist gas, which regulates the amount of high-temperature oxidation occurring in the hole. As the workpiece thickness decreases, the lens/workpiece distance required to achieve the best surface condition or the least amount of dross also decreases. For workpiece thicknesses less than 0.3mm, the lens/workpiece distance is greater than the lens focal length. These observations are verified by an experimental study of the effects of changing parameters on surface debris, dross formation, and hole taper [60].

For laser percussion-drilling processes on metals, drilling efficiency is strongly dependent on the phenomenon of laser-supported detonation waves [47]. Plasma formation at the workpiece surface due to laser/material interaction severely decreases laser drilling effectiveness because it absorbs a significant portion of the incoming beam energy and thus shielding the workpiece surface. A laser-supported detonation (LSD) wave occurs when the absorbing plasma layer is coupled to a shock wave occurring when material at the erosion front detonates. This transient phenomenon propagates the plasma volume and effectively shields the workpiece surface from the laser beam. LSD wave effects can be minimized by using a shorter focal length lens, by increasing the beam intensity and by using a low density gas jet such as He instead of air. By minimizing the LSD wave effect, holes up to 65mm in depth can be drilled in copper workpieces using energy densities between $300J/cm^2$ and $6000J/cm^2$.

One major application for laser drilling is making holes for combustors and turbines in aircraft jet engines. The holes are used for cooling purposes, where a cool air stream is either mixed with the hot combustion gases or directed to maintain the flame tube and turbine blade surfaces at a desired temperature. A typical combustor has approximately 30,000 holes of less than 1.5mm diameter (Fig. 6.3). Since the combustor and turbine are usually made from tough heat-resistant alloys, mechanical drilling methods are both expensive and time-consuming. Therefore, aerospace manufacturers have turned to non-traditional machining methods, such as laser machining, electro-chemical machining, electro-discharge machining, and electron beam machining. Drilling experiments were performed on combustors using Ruby, Nd:Glass, and Nd:YAG lasers [17]. The lasers were operated in a pulsed mode with pulse frequencies less than 10Hz and pulse energy between 10J and 40J. The results showed a degree of asymmetry in the hole profile, a significant layer of resolidified material, and formation of microcracks near the hole. A trepanning method, where the laser beam is moved in a circular motion along the plane of the workpiece to produce a hole, can also be used in conjunction with a pulsed Nd:YAG beam operating at 300Hz. The trepanning method produces holes with significantly reduced recast layers, less microcracking, and less hole taper. Additionally, by controlling the relative motion between the workpiece and beam during trepanning, non-circular hole geometries can also be produced.

Material	Laser Type	Laser Power/ Power Density	Drilling Speed	Hole Depth	Hole Diameter	Refs
Aluminum	CO_2	$800kW/cm^2$		0.78mm		18
Brass	CO_2	$800kW/cm^2$		0.78mm		18
Copper	Nd:YAG	$2MW/cm^2$		0.045mm	0.25mm	47
Nickel	Nd:YAG	10-20J, 1-3.5ms	1ms per hole	0.5-1.25mm		61
Steel	Nd:YAG	8kW	0.06ms per hole	0.8mm	0.2mm	50
St. Steel	CO_2	$800kW/cm^2$		0.61mm		18
Tantalum	Nd:YAG	10-20J, 1-3.5ms	1ms/hole	0.5-1.25mm		61
Titanium	Nd:YAG	10-20J, 1-3.5ms	0.67ms per hole	0.5-1.25mm		61

TABLE 6.2   Representative Data for Laser Drilling of Metals

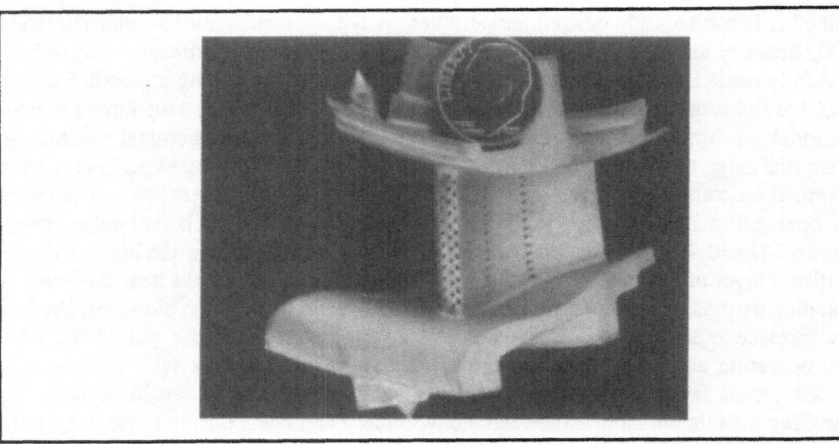

FIGURE 6.3   Laser-Drilled Guide Vane for Aircraft Turbine [14]

6.2.2   Two-Dimensional Machining: Cutting

Laser cutting of metals has been successfully applied in industry to cut two-dimensional shapes on workpieces with thicknesses up to 10mm (Table 6.3). Lasers have also been applied for cutting intricate contours on curved workpieces in three dimensions [14]. Two types of laser cutting can be used for processing metals: *reactive gas cutting* and *laser fusion cutting*. For reactive cutting of metals, an O_2 gas jet is used, and material removal is achieved through high-temperature oxidation reactions. The laser beam serves as a high-temperature heat source to propagate chemical reactions. In fusion cutting of metals (Fig. 6.4), an inert gas jet is used while the laser beam serves as a heating source to melt material [39]. In general, higher material removal rates can be achieved through reactive gas cutting; however, fusion cutting achieves better surface quality and dimensional accuracy of the kerf. Some types of metals such as copper and tungsten have high reflectivity at the wavelength for CO_2 laser radiation and are difficult to cut using a CO_2 laser [4]; in order to circumvent this problem, an absorbant coating can be applied to the workpiece surface.

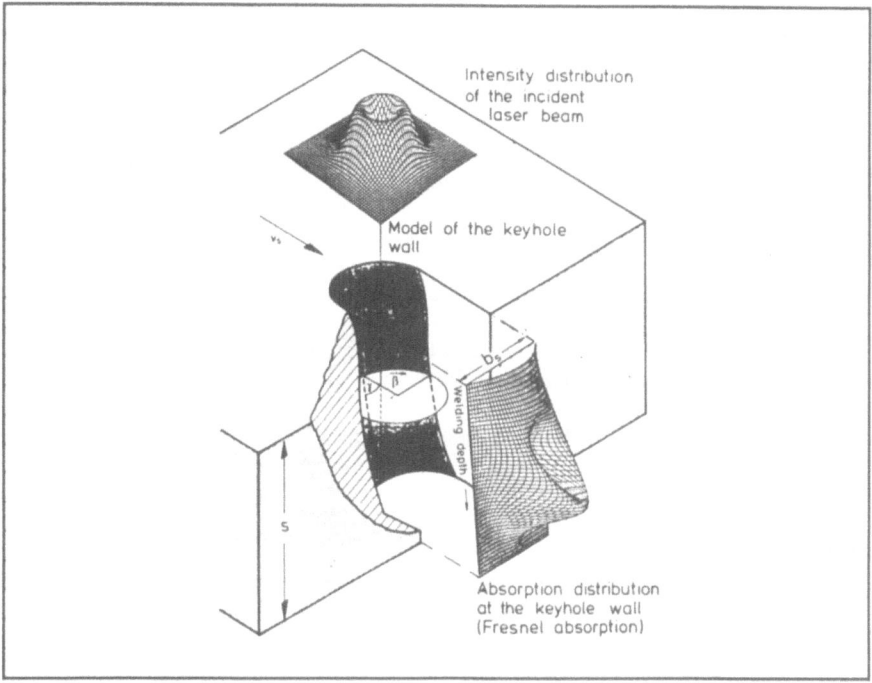

FIGURE 6.4   Schematic of Cutting Front for Laser Fusion Cutting of Metals [39]

In [44], reactive gas laser cutting was performed with an oxygen jet. During the oxy-laser cutting process, the kerf zone was heated up to ignition by the laser beam while the oxygen jet burns the material and blows the slag off. These experiments were performed on mild steel and austenitic Cr-Ni steel workpieces using a 5 kW laser operating in a TEM_{04} mode. The laser beam propagated through a telescopic system which was installed to reach optimized focusing conditions over moving mirrors to a moving laser nozzle. A triple jet nozzle is used to achieve a coaxial laser-beam-cutting jet arrangement with a supersonic gas jet. This nozzle design is necessary in order to achieve high surface quality in thicker workpieces. For the cutting experiments, optimum cutting speed and power were found for producing kerfs with minimal surface roughness for mild steel 4mm to 35mm thick and high alloyed austenitic steel 4mm to 15mm thick. As the plate thickness decreases, the cutting speed increases and kerf width decreases. Moreover, higher cutting speeds combined with a narrower focal spot were found to cause lower thermal influence and a smaller distortion near the kerf. Oxygen-assisted laser cutting is not as effective on high-Nickel alloy steels as with other types of steel. One possible cause of this loss in effectiveness is the large temperature increase at the erosion front during the laser cutting process, which leads to the formation of nickel and chrome oxides. The accumulation of molten oxides at the erosion front, especially in deep kerfs, reduces the laser cutting effectiveness for workpiece thickness beyond 15 mm.

An analysis of the erosion front for laser fusion cutting of metallic workpieces (Fig. 6.4) shows phase change as the primary mode of material removal [39]. The absorption of laser beam energy on the erosion front depends on the inclination of the erosion front relative to the laser beam direction.

An investigation of laser fusion cutting of thin steel and aluminum workpieces (less than 10mm in thickness) was performed in [42] with a 1 kW CW CO_2 laser (Fig. 6.5). In general, laser cutting of thin materials attained the best quality and rate when the cutting vaporized the material. The 1kW beam easily provided the energy density required for material vaporization. Based on the results in [42], cutting rates for metallic and non-metallic thin materials were found to be:

- Inversely proportional to the thickness of the material
- Proportional to the input power level beyond a certain threshold value

For workpieces less than 10mm thick, beam divergence is minimal and beam energy can be concentrated in a focussed spot. This results in higher energy densities and higher material removal rates than those achievable with workpieces more than 10mm thick. Also, the molten material can be expelled easily from thin workpieces. As a result, changes in cutting depth are roughly proportional to changes in energy density. For thick workpieces, the cutting depth may exceed the working distance of the focussing lens,

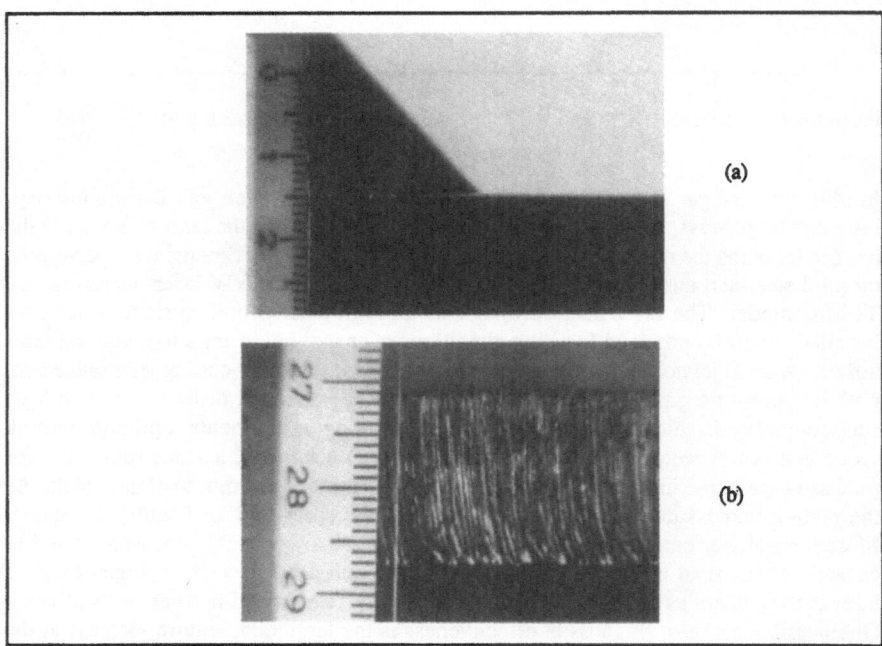

(a)

(b)

FIGURE 6.5   Laser Fusion Cutting of (a) 9mm and (b) 16mm Steel [43]

while the molten material is difficult to expel; consequently, a fraction of the beam energy may be dissipated or absorbed by the molten layer, reducing cutting efficiency.

The use of reactive gases increases cutting speeds in laser cutting processes that are dominated by chemical reactions; however, the use of non-reactive gases produces kerfs with higher surface quality. The self-burning of the material, can lead to kerf quality problems and eventually defines the lower limits of the cutting rates. In [19], laser cutting was analyzed for metallic workpieces such as stainless steel (Fig. 6.6), tool steel (Figs. 6.7 and 6.8), armor plate, titanium alloy (Fig. 6.9) and aluminum (Fig. 6.10) with thicknesses between 10mm and 15mm.

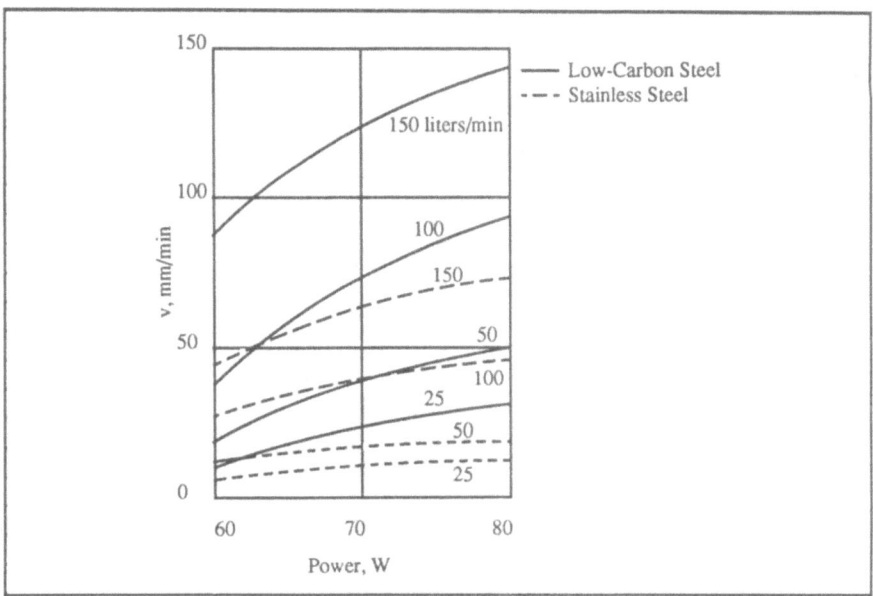

FIGURE 6.6   Cutting Speed vs. Laser Power for 1mm Thick Stainless Steel and Low Carbon Steel With Different Oxygen Flowrates [4]

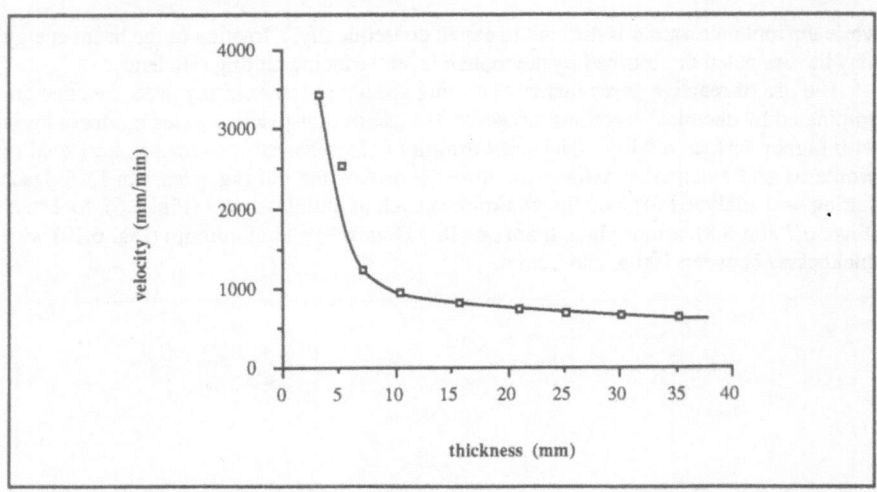

FIGURE 6.7   Cutting Speed vs. Depth of Cut for Mild Steel With Laser Power of 4kW [44]

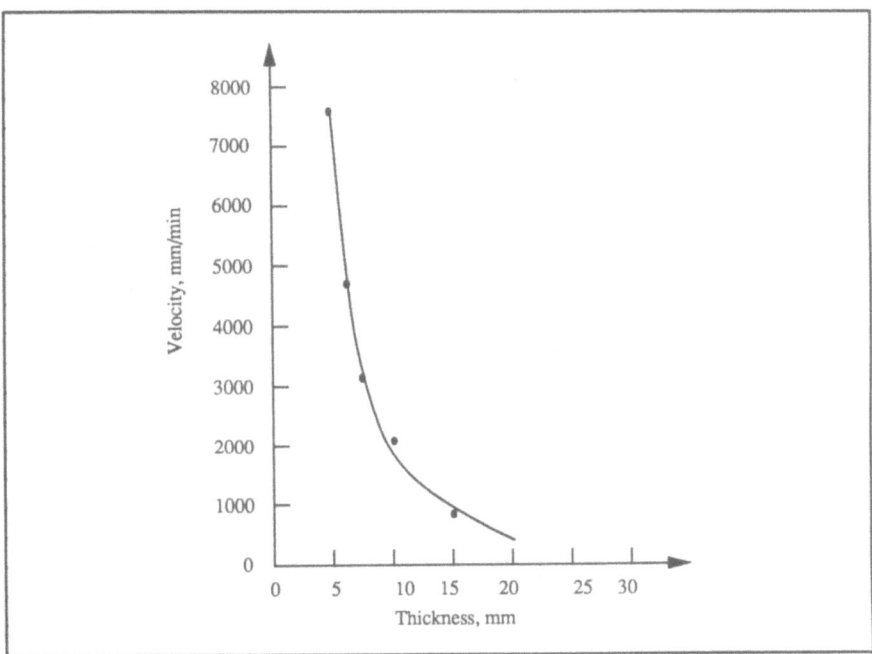

FIGURE 6.8   Cutting Speed vs. Depth of Cut for Austenitic Steel With Laser Power of 4kW [44]

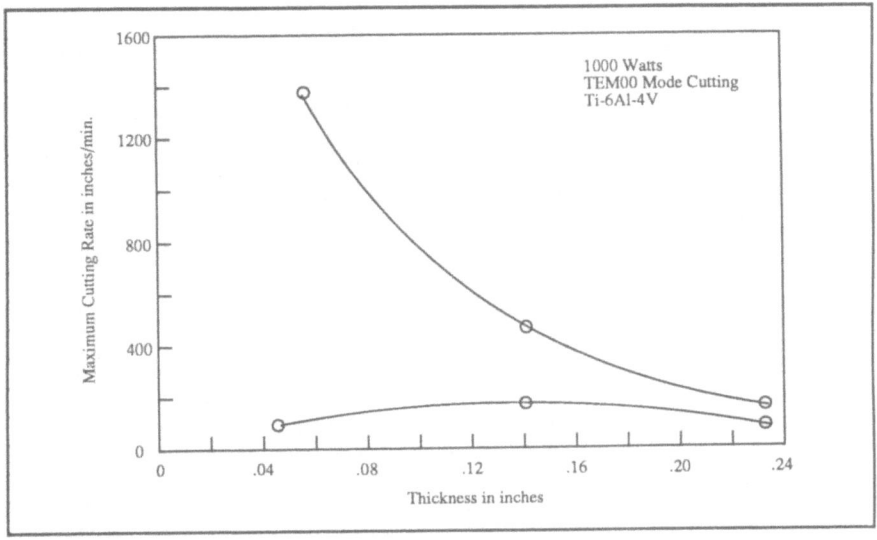

FIGURE 6.9   Cutting Speed vs. Depth of Cut for Titanium Alloy [20]

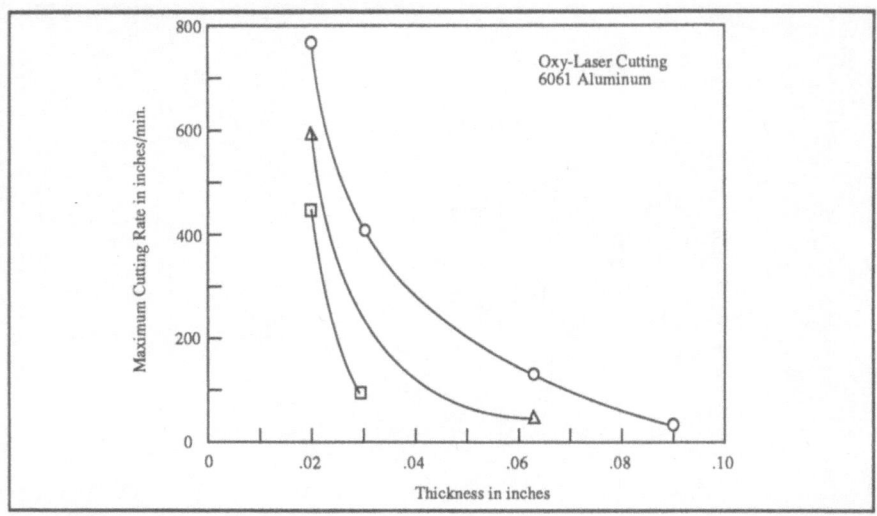

FIGURE 6.10   Cutting Speed vs. Depth of Cut for Aluminum [20]

In laser cutting of metals, several factors can affect cutting surface quality. In [3] the relationship between kerf quality and cutting parameters was investigated for laser-gas-cutting of stainless steels. Using a coaxial oxygen jet arranged with a focused laser beam to cut stainless steel, the resulting surface quality was relatively low due to the formation of an oxide "dross" clinging to the bottom. In order to improve this condition, two methods, *pile cutting* and *tandem nozzle cutting*, can be used. Pile cutting entails placing a thin sheet of mild steel on top of the stainless steel workpiece. Tandem nozzle cutting uses an off-axial rear gas jet along with the coaxial nozzle. Both techniques improve the surface roughness and flatness of the kerf. Pile cutting can produce a dross-free kerf, since molten oxide reduces the viscosity of the melting flow on the cutting front and promotes separation of the molten oxide dross. Tandem nozzle cutting can be employed with two jet nozzles, resulting in an enhancement of the gas jet force and increased removal of the molten metal and its oxides.

In [34], different assist gases were used to analyze the effects on dross formation. When stainless steel material is cut using an oxygen assist gas, the dross formation is mainly composed of iron oxides and chromium oxides. Due to its high melting point, the dross material resolidifies before the gas jet can expel it completely from the kerf. This dross formation can be reduced by using a gas mixture (60% CO_2 and 40% O_2), which reduces the formation of oxides and helps in ejecting the molten material. Furthermore, a mixed gas also improves the weldability of the cutting kerf surface.

Another issue in surface quality is the formation of striations along the kerf walls. In a study on striations [30], two distinct zones on the kerf edge were found. Zone I, the zone of regular striations near the kerf entrance, is thought to be the result of oxygen-assist laser beam heating. Zone II, the zone of indistinct striations near the kerf bottom, is believed to be caused by a diffusive thermochemical reaction in the absence of direct contact with the laser beam. In [30], an increase of Zone I depth was shown to correspond with an increase in beam pulse-width. The surface roughness was found to depend on pulse-width in the vicinity of the top edge approximately 0.2mm to 0.8mm from the kerf entrance.

In [4], the microstructure of laser-cut kerfs for low-carbon and stainless steel workpieces up to 10mm thick was studied. Due to the high temperatures reached at the erosion front, the material near it undergoes phase transformation, resulting in the creation of a heat-affected zone. While the laser beam energy is not sufficient to melt the material in the heat-affected zone, it is high enough to heat-treat the material to a harder state. The results of [4] showed that the heat-affected zone increased with workpiece thickness. The width of the heat-affected zone on the beam entry side was also found to be less than that on the exit side. The change in microhardness of the kerf edges was greater on the exit side, but the microhardness was normal at a distance of less than 0.1 mm from the edge of the kerf. Because the laser beam energy is concentrated into a small spot, the phase transformation effect is highly localized and the heat-affected zone is small.

The efficiency of laser cutting for other metals such as aluminum and copper depends on surface finish as well as workpiece thickness. Since these materials are highly reflective, sheets with smooth untreated surfaces cannot be easily cut with a laser. The cutting efficiency can be improved by either creating a rough surface to absorb both incident and reflected laser beam energy, or by using an absorbant coating to engage the incoming laser beam.

Material	Laser Type	Laser Power/ Energy Density	Cutting Speed	Kerf Depth	Kerf Width	Refs
Aluminum	Nd:YAG	200W	0.23m/min	1.3mm	0.6mm	30
Aluminum	Nd:YAG	1.25kW	1.5m/min	3mm		39
Aluminum	Nd:YAG	200W	0.78m/min	2mm	0.25mm	55
Copper	Nd:YAG	300-6000J/cm^2		6.5mm		47
Copper	Nd:YAG	120W	0.05m/min	3mm	0.2mm	55
Gold	Nd:YAG	190W	0.015m/ min	3mm	0.35mm	55
Hastalloy	Nd:YAG	120W	0.45m/min	2mm	0.2mm	55
Hastalloy	Nd:YAG	120W	0.02m/min	8mm	0.2mm	55
Mild Steel	CO_2	100W	1.6m/min	1mm		4
Mild Steel	CO_2	250W	0.635m/ min	0.5mm		4
Mild Steel	CO_2	850W	1.8m/min	2.2mm		4
Mild Steel	CO_2	1kW	0.8m/min	10mm	0.9mm	44
Mild Steel	CO_2	2.5kW	0.85m/min	10mm	0.9mm	44
Mild Steel	CO_2	4kW	0.95m/min	10mm	0.9mm	44
Mild Steel	Nd:YAG	150W	0.23m/min	1.5mm	0.6mm	30
St. Steel	CO_2	100-850W	0.94-2.6m/min	1-9mm		4
St. Steel	Nd:YAG	1.15-1.2kW	0.45-1.5m/min	3-6mm		39
Titanium	CO_2	250-850W	0.2-3.24m/min	0.5-0.6mm		4
Titanium	CO_2	500-1000W	15-30in/min	0.06-0.125in		20
Titanium	CO_2	375W	100-360in/min	0.06-0.125in	0.03in	14
Tungsten	Nd:YAG	180W	0.03m/min	2mm	0.3mm	55

TABLE 6.3　　　Representative Data for Laser Cutting of Metals

6.2.3 Two-Dimensional Machining: Scribing and Marking

Laser scribing and marking on metals has been shown to be economically viable for some engineering applications as well as for part identification in industry (Table 6.4). Laser marking systems are currently employed in the electronics, cosmetics, food and beverage, and optical industries, among others. Engineering applications include the grooving of micro-channels for cooling systems and the creation of slots for assembly. The engraving of products with alphanumeric or bar codes, logos, symbols and graphics are also useful applications for CO_2 and Nd:YAG lasers. Laser marking (Fig. 6.11) compares favorably with other marking systems, when the comparison is based on throughput, performance and flexibility (Fig. 6.12).

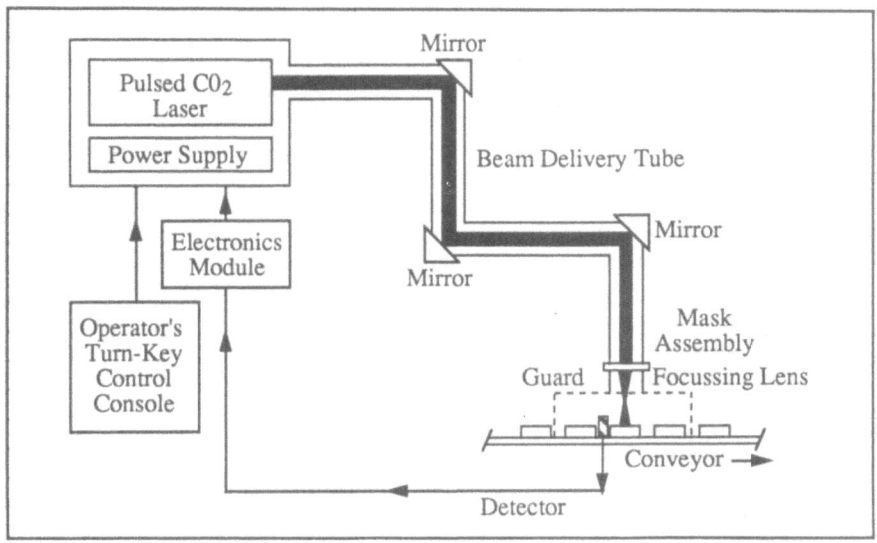

FIGURE 6.11   Setup for Laser Marking Equipment [25]

	Permanence	High Throughput	Rapid Text Change	Comments
Metal Stamp	X	X		High consumables cost
Ink Jet		X	X	High maintenance time
Labels/ Nameplates		X	X	Added production process
Pantograph (pencil)	X		X	High consumables cost
Silkscreen		X		High consumables cost
Molded/ Embossed	X	X		High consumables cost
Chemical Etch	X			Acid disposal
Laser Marking	X	X	X	

FIGURE 6.12   Comparison of Laser vs. Conventional Marking [23]

The basic requirements for a CO_2 laser marking system include a Transversely Excited Laser (TEA), power supply, modulator, gas jet, motor control electronics, two beams including folding mirrors and imaging lenses, and a detection system to guide the beam scanning direction [25]. The laser generates a high peak energy which vaporizes the surface of most metals and creates deep, clean scribe lines to form patterns on a metallic surface.

One typical industrial CO_2 laser marking application can be found in the pharmaceutical industry, where lasers are used to mark statutory information such as manufacture and expiration dates on the crimped end of collapsible creams tubes. The use of laser marking negates the difficulty associated with preprinting. Laser marking is also used in the distillery industry, where lasers apply permanent security codes to the side of the bottle.

Nd:YAG laser marking systems usually operate in one of two modes: dot matrix and continuous lines [23]. Each functions differently and each is suited for specific applications. The dot matrix systems utilize a pulsed Nd:YAG laser in conjunction with a raster scanning beam delivery system. Dot matrix marking is effective for relatively large length scales (greater than 1mm), but is not suitable for small lettering applications, where the distance between dots approaches the maximum excursion of the line. Continuous line marking systems, which produce legible alphanumeric characters, are the most commonly used laser markers. With continuous marking, small lettering or fancy scripts can be created (Fig. 6.13). The achievable groove depths depend primarily on the workpiece material and the energy density as applied to the workpiece (Figs. 6.14 and 6.15). Some representative applications for YAG laser engraving include marking nameplates, bar code marking and medical applications. Future expansion of laser engraving will significantly depend on developments in computer motion control, imaging systems and robotics.

Material	Laser Type	Laser Power	Groove Depth	Refs
Mild Steel	CO_2	1-10MJ/cm^3	0.3-2mm	7
St. Steel	CO_2	1-10MJ/cm^3	1.6-4.5mm	7

TABLE 6.4   Representative Data for Laser Scribing of Metals

FIGURE 6.13   Laser-Marked Tool Bit [23]

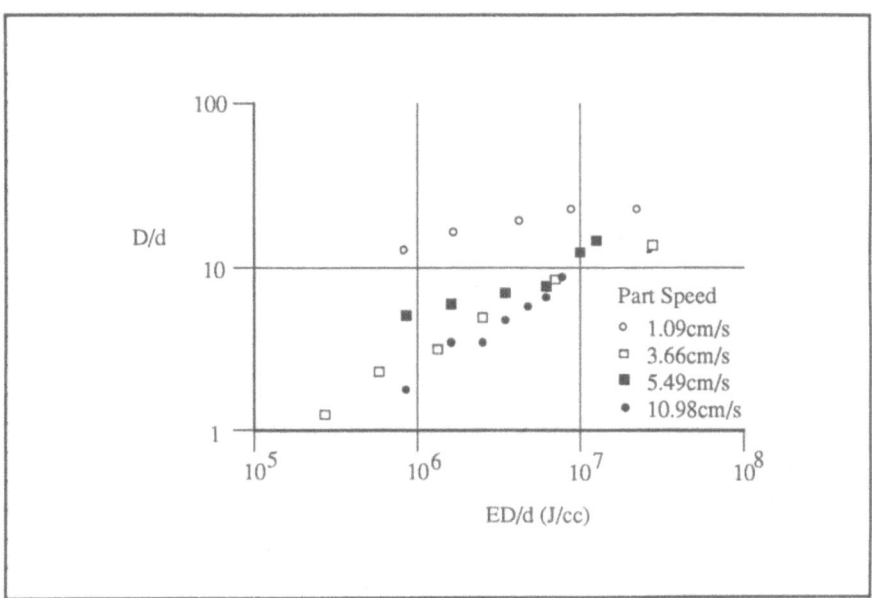

FIGURE 6.14   Groove Depth vs. Energy Density for Mild Steel

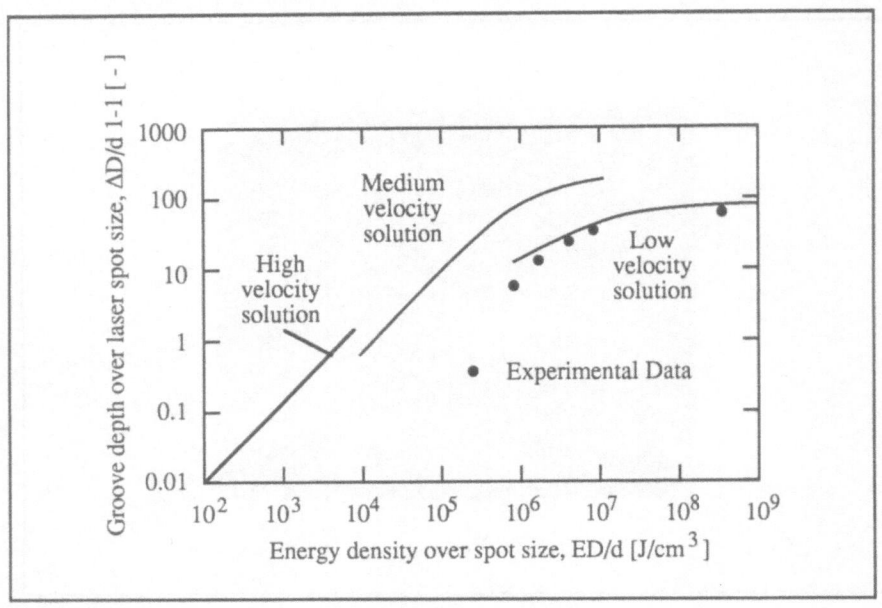

FIGURE 6.15   Groove Depth vs. Energy Density for Stainless Steel

6.2.4   Three-Dimensional Machining

One technique for producing three-dimensional shapes which is particularly applicable to steel combines two-dimensional laser cutting with laser welding to construct a three-dimensional part from two-dimensional shapes. In this technique, two-dimensional shapes are first cut by one beam, and are then butt-welded together to form one part. In [5, 35], a system composed of two 500W CO_2 lasers was used in conjunction with a 5-axis NC positioning system. Helium and nitrogen cover gas were used for both cutting and welding. The test results in [35] showed that cutting speeds of up to 10.5m/min were achievable for workpiece thicknesses up to 0.8mm.

For laser turning, threading and milling (Fig. 6.16), two laser beams can be used to produce converging grooves on a workpiece [7]. When the two grooves intersect, a volume of material is removed. This technique offers higher energy efficiency, higher material removal rates and more flexibility than non-traditional machining processes. Since most of the material is removed in chip form, the energy required to remove the material is significantly less than that used in single-beam melting of the material. Also, since material is removed chunk by chunk instead of molecule by molecule, the time required to remove the material is greatly reduced. Finally, the use of two beams results in a great deal of flexibility in terms of part shapes that can be produced.

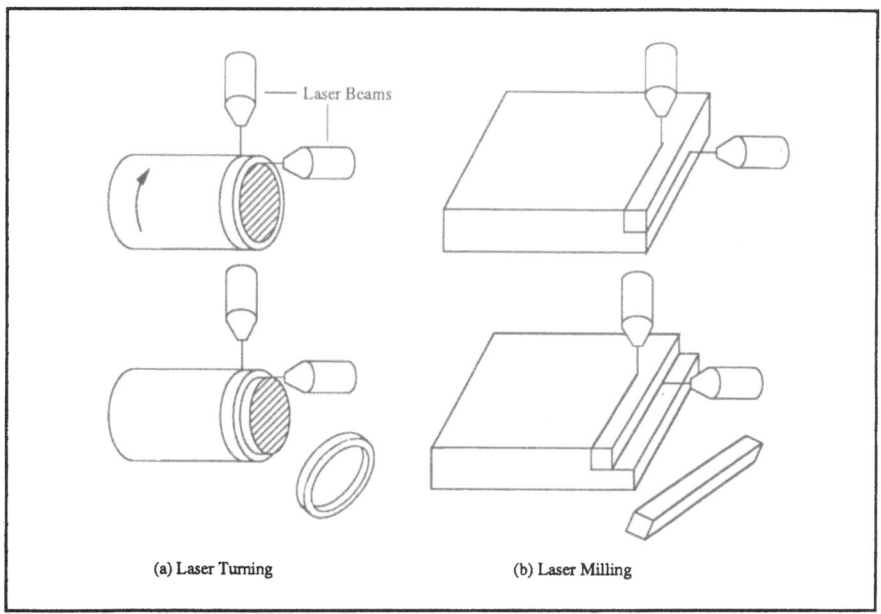

FIGURE 6.16   Schematic for (a) Laser Turning and (b) Laser Milling for Steel [7]

In laser turning experiments performed on 5cm diameter cylindrical stainless steel workpieces [7], energy densities above $10^6 J/cm^2$ were required to form chips with a cross-sectional area of $1cm^2$. For a beam power of 500W, over 30 beam passes across the workpiece were required to remove material. In some cases, the chip to be removed was joined onto the workpiece surface by resolidified molten material; this problem was minimized by using an off-axis gas jet to eject molten material from each groove. Use of off-axis gas jets also improved the surface quality on the finished part by minimizing dross formation. In general, the use of reactive gas jets results in larger chip volumes removed, but the resulting surface quality is poor and the depth of cut cannot be easily regulated. Use of inert gas jets produce smaller chip volumes, but the surface quality of the finished part is improved and variations in the depth of cut for each groove can be minimized. Based on the experimental results one can also conclude that effective laser turning, milling and threading of metals and particularly steel would require higher power (around 2-3 kW in the CW mode).

6.3   Ceramics

6.3.1   One-Dimensional Machining: Drilling

Lasers can be used very successfully in machining of ceramic materials. Due to their high hardness and brittleness, conventional machining of ceramics by mechanical or ultrasonic methods is costly and time-consuming. For example, ultrasonic drilling of aluminum oxide requires approximately 30 seconds per hole [24]; furthermore, the cost of ultrasonic drilling is approximately three times that for laser drilling. Mechanical drilling requires additional operations such as attaching the workpiece to a fixture with adhesives prior to drilling, and a cleaning operation after drilling. Laser drilling can be a more time- and cost-efficient process. Laser machining can be applied either before or after "firing" of the ceramic material. The data and results presented in this section primarily refer to machining of ceramic workpieces after firing of the substrate or sintering (Table 6.5). However, some drilling operations have also been performed on "green",.(unfired ceramic workpieces), substrate materials [24]. Two major problems exist with laser machining of green substrates. First, since laser machining produces localized heating, the material near the rim of the hole becomes brittle and may break off. Second, during the firing process, anisotropic shrinkage of up to 20% may occur, causing large deviations in dimensional accuracy of the machined part.

Like metals, ceramic materials can be drilled either by percussion (Fig. 6.17) or trepanning (Fig. 6.18) methods on materials such as aluminum oxide (Fig. 6.19).

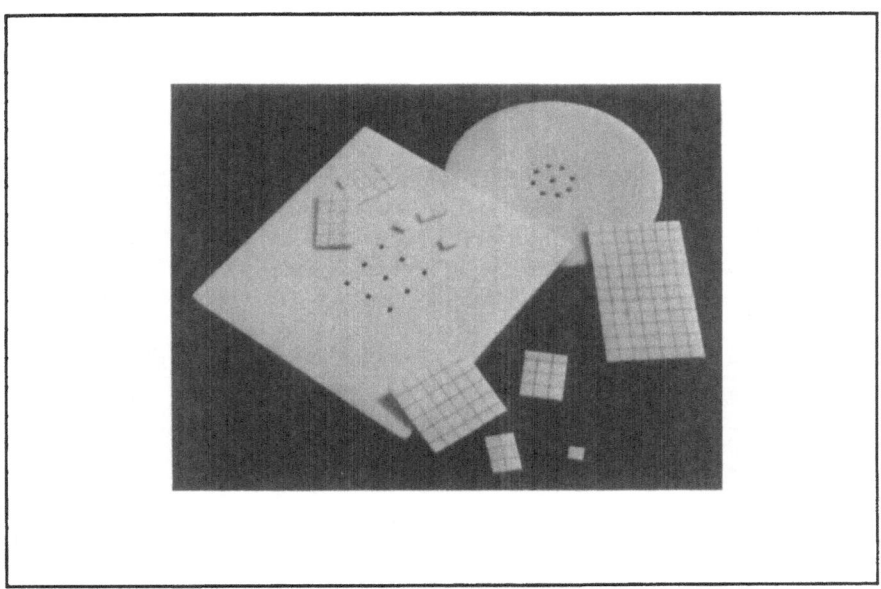

FIGURE 6.17   Laser-Drilled Alumina Substrate [29]

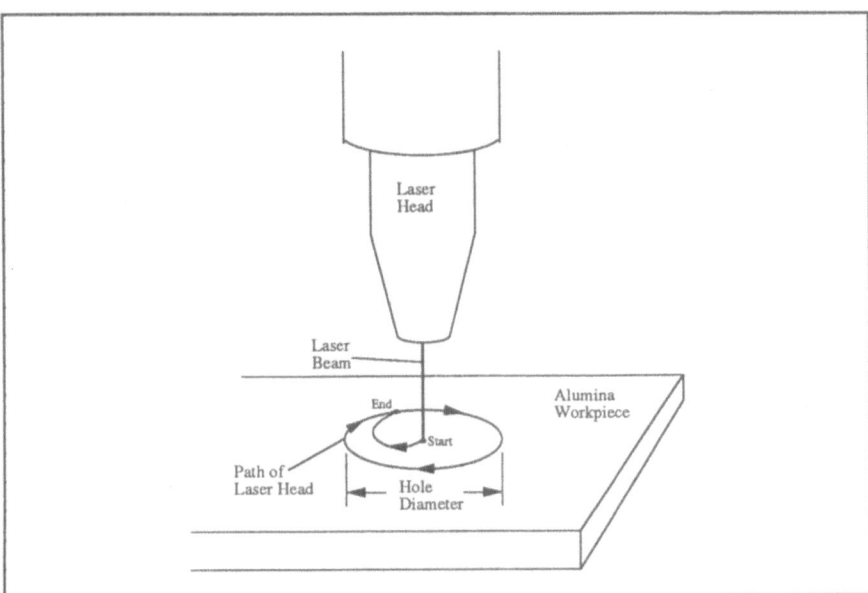

FIGURE 6.18   Laser Drilling by Trepanning on Aluminum Oxide [29]

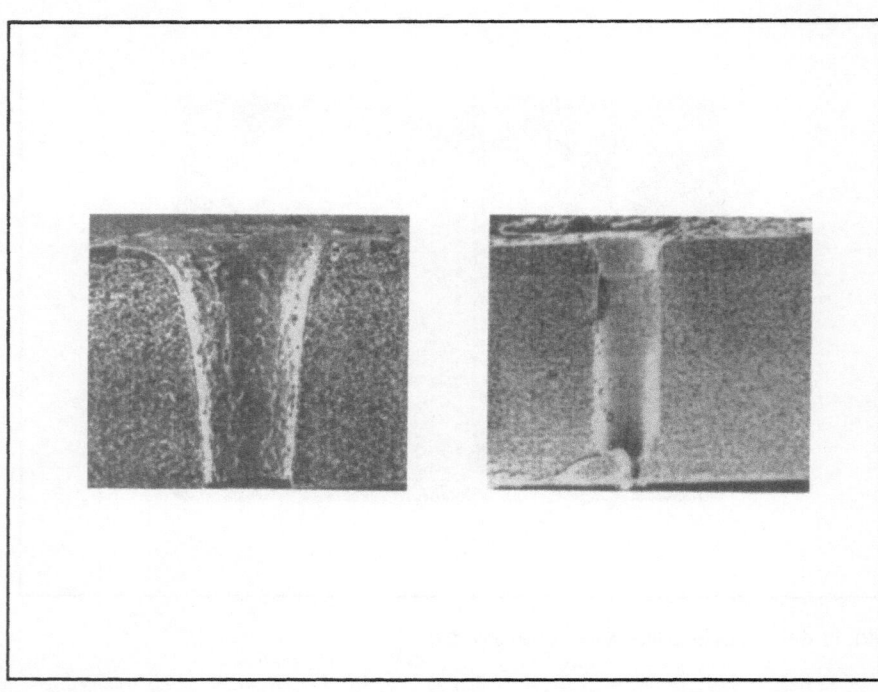

FIGURE 6.19   Comparison of Pulsed CO_2 and Pulsed Nd:YAG Laser-Drilled Holes in Aluminum
               Oxide [24]

One consideration in laser drilling of ceramics is the energy required to initiate drilling. A
threshold energy density value is defined as the energy density level below which
material removal is not possible. For uncoated aluminum oxide [56], the threshold
energy density ranges from 750J/cm^2 to 1000J/cm^2. This energy density threshold can
be decreased to approximately 400J/cm^2 by coating the top surface of the workpiece with
gold, which has a relatively high thermal conductivity. In [6], a threshold energy density
value of 2000J/cm^2 was found for non-porous aluminum oxide (Fig. 6.20).

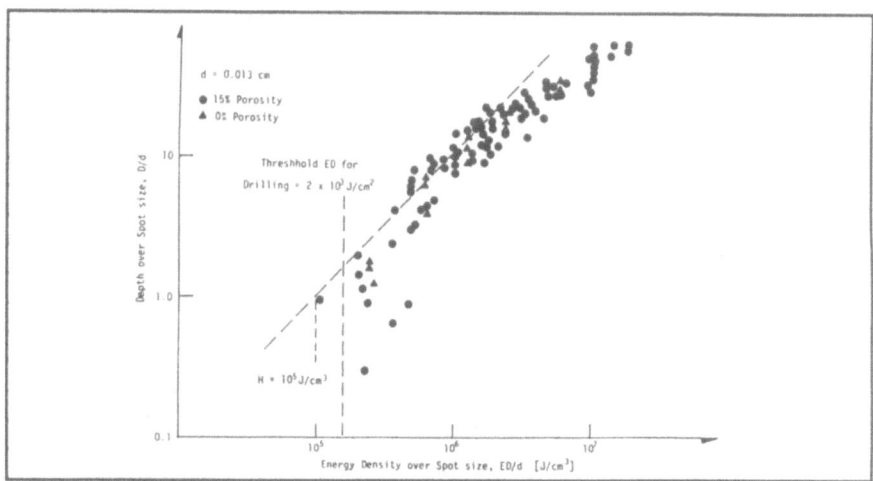

FIGURE 6.20   Drilling Depth vs. Energy Density for CO_2 Laser Drilling of Alumina [6]

In [56], holes up to 550μm deep were drilled with energy densities of up to 7000J/cm^2 using a pulsed ruby laser. The hole shape was found to be roughly proportional to the spatial beam intensity profile. Drilling of 0.25mm diameter holes in a 0.1mm thick workpiece using a pulsed CO_2 laser beam was performed in [14]. The pulse frequency was 500Hz and pulse duration was 200ms. A machining speed of 0.1 seconds per hole was achieved, and a dimensional accuracy of ±0.02mm for hole position and ±0.05mm for hole diameter variation was obtained. A 1.2kW CW CO_2 laser was used to drill blind holes in aluminum oxide [6] with hole depths ranging from 0.02cm to 1cm (Fig. 6.20). Energy density values ranged between 2kJ/cm^2 and 500kJ/cm^2. In [24], holes of up to 0.25mm in diameter were drilled in aluminum oxide with power densities between 10^{6}W/cm^2 and 10^{8}W/cm^2 using CO_2 and Nd:YAG lasers. Drilling time was less than 0.2 seconds per hole. CO_2 laser-drilled holes showed a noticeable taper, while the Nd:YAG laser-drilled holes had straight edges. Since the Nd:YAG laser beam can be focussed to a smaller spot size than the CO_2 beam, the working distance for a focussed Nd:YAG laser beam is larger and the beam divergence effect is less significant than for a CO_2 laser, resulting in straight holes.

One problem encountered in ceramic drilling is the resolidified material which forms in the hole. Since the melting temperatures for ceramic materials are high, material removal occurs primarily through melting and a molten layer is formed at the erosion front. Due to the limited space in the hole, expulsion of molten material with a gas jet is difficult. The remaining recast layer reduces hole depth by absorbing beam energy and degrades surface quality by increasing hole surface roughness. In [29], this problem was addressed by coating the substrate with polyvinyl alcohol prior to drilling. The coating minimized adhesion of recast material around the hole entrance. The recast layer inside the hole was removed by a secondary glass bead peening process.

Micro-crack formation near the drilled hole is another problem encountered in laser drilling of ceramics (Fig. 6.21). Cracks are induced by rapid heating during laser

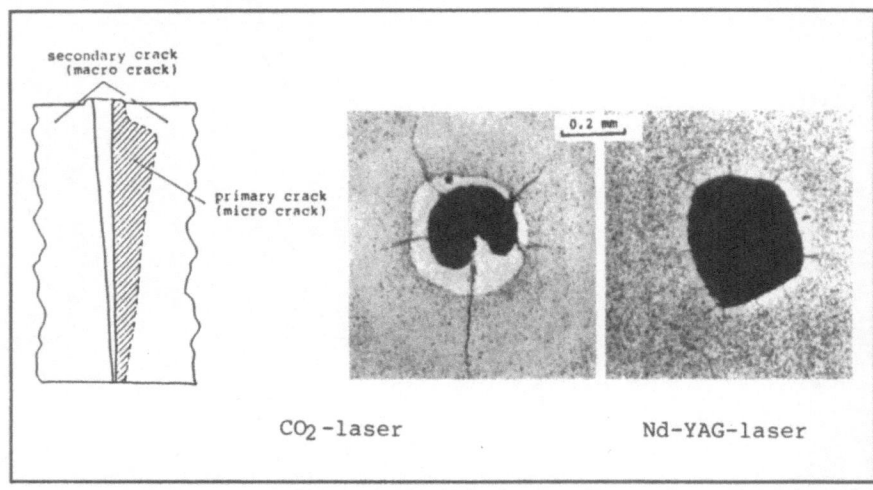

FIGURE 6.21   Crack Formation in CO_2 and Nd:YAG Laser-Drilled Holes [26]

processing, which causes large temperature gradients near the hole surface. Microcracking can be minimized by preheating the workpiece to an elevated temperature prior to laser machining. In this manner, the temperature gradients achieved during laser machining will be reduced significantly compared to laser drilling of a cold workpiece.

An industrial application of laser drilling of ceramics is the manufacturing of distributor plates for fluidized bed heat exchangers [40]. The distributor plate is used to support the bed material and ensure uniform distribution of the gas over the plate. In many applications, the gas is at a very high temperature (1100-1700°C) and is corrosive, so ceramics such as silicon carbide, silicon nitride and aluminum oxide are the only materials applicable. Each distributor plate must have 3% to 10% of the surface area open to allow gas passage, so an array of holes with less than 1.5mm diameters at a density of approximately 1500 holes per square foot must be drilled. This drilling process is very difficult with conventional techniques but is well-suited for laser machining. In [40], a 400W Nd:YAG laser was used with pulse frequencies of up to 200Hz and pulse durations between 250ms and 1000ms. Silicon carbide, silicon nitride, aluminum oxide, and COMPGLAS (silicon carbide fibers in a lithium aluminum silicate glass matrix) workpieces of 3mm to 3.5mm in thickness were used. Holes from 0.25mm to 1.5mm in diameter were drilled at a rate of two to three seconds per hole. Silicon carbide generally required the highest pulse energy and formed the most irregular holes. Slots of 1.5mm in length were also produced using beam trepanning, which produced machining rates of 20 seconds per slot.

Material	Laser Type	Laser Power/ Power Density	Drilling Rate	Hole Depth	Hole Diameter	Application	Refs
Aluminum Oxide	Ruby	ED=7kJ/cm^2	6ms/hole	550µm		Percussion Drilling	56
Aluminum Oxide	CO_2	150W	0.1s/hole	0.025in	0.01in	Percussion Drilling	14
Aluminum Oxide	CO_2	1.2kW		0.02-1cm	0.03cm	Percussion Drilling	6
Aluminum Oxide	CO_2	1kW	44ms/hole	0.027in			29
Aluminum Oxide	Nd:YAG	100MW/cm^2	0.2s/hole	0.63mm	0.25mm	Percussion Drilling	24
COMP GLAS	Nd:YAG	400W	2-3s/hole	3.5mm	0.25-1.5mm		4
Silicon Carbide	Nd:YAG	400W	20s/slot	3mm	1.5mm	Trepanning	4
Silicon Carbide	Nd:YAG	400W	2-3s/hole	3.5mm	0.25-1.5mm	Fluidized Bed Heat Exchanger	4
Silicon Nitride	Nd:YAG	400W	2-3s/hole	3.5mm	0.25-1.5mm		4
Silicon Nitride	Nd:YAG	400W		5mm	0.3mm		26
Silicon Nitride	CO_2	1.5kW		4.5mm	0.3mm		26

TABLE 6.5   Representative Data for Laser Drilling of Ceramics

6.3.2 Two-Dimensional Machining: Cutting

Conventional cutting of ceramic materials is usually done with a diamond saw. Although this gives a high surface quality, the cutting speed is low, typically around 20mm/min [24]. Laser cutting allows cutting speeds up to 1200mm/min, a significant increase over conventional machining, without sacrificing surface quality (Table 6.6).

In [41], laser through-cutting was performed on ceramics with thicknesses from 0.6mm to 4.0mm, laser power from 150W to 500W, and cutting speeds that ranged from 0.1m/min to 1.2m/min (Fig. 6.22). Material removal occured primarily through local vaporization of the material. In [1], aluminum oxide material up to 15mm thick was cut with a 10kW Nd:YAG laser and a cutting speed from 10m/min to 20m/min. The resulting kerf surface had a roughness of less than 2µm RMS. The best results (small taper, smooth surface, minimal micro-cracking, and large cutting depth) could be achieved by using high pulse energies and a high pressure assist gas (a 600kPa N_2 jet). By using the same laser conditions, silicon carbide workpieces up to 5mm thick could be cut with a speed of 40mm/min, 3mm thick silicon nitride material could be cut at

50mm/min, and 1.3mm thick zirconium oxide could be machined at 200mm/min. In [24], cutting of alumina material was performed with a CO_2 laser at power densities from $10^6 W/cm^2$ to $10^8 W/cm^2$. Cutting speeds of 3mm/s to 5mm/s were achieved. Material in the kerf was melted and either ejected by a gas jet or partially evaporated.

One method for minimizing micro-crack formation is to pre-heat the workpiece to a high temperature prior to interaction with the laser beam. This procedure minimizes the temperature gradients and thermal stresses in the workpiece during laser machining. Laser cutting of silicon nitride, silicon carbide, and magnesia-stabilized zirconia (PSZ) workpieces were also investigated [21] using a 15 kW CW CO_2 laser with a minimum spot diameter of 2.7mm. To minimize cracking, the specimens were heated first in a furnace to 1400°C. A gas jet was used to reduce oxidation and prevent plasma formation. Silicon nitride requires a heating and cooling sequence to avoid fracture; the specimens were first heated to 1000°C in 200°C steps, then machined and allowed to cool to ambient temperature. Beam power ranged from 1 to 4 kW with cutting speeds ranging from 15mm/min to 150mm/min. For the experiments in which one beam pass was made, the specimen did not fracture. The width of the groove ranged from 0.66mm to 4mm at the top and from 0mm to 2.3mm at the bottom. The depth ranged from 6.35μm to 740μm. The area from which material was removed ranged between 5.8mm^2 and 1500mm^2 and the material removal rate was between 33mm^3/min and 2300mm^3/min, an increase in the cutting speed resulted in a reduction in MRR. Silicon carbide behaved similarly to silicon nitride. PSZ was machined at 1225°F and 1825°F with single and multiple passes. The beam power ranged from 2kW to 6kW and the cutting speed varied from 9100 to 16000mm/min. The resulting depth of cut for PSZ ranged from 0.13mm to 0.9mm. In general, higher beam power and preheat temperature resulted in higher material removal rates.

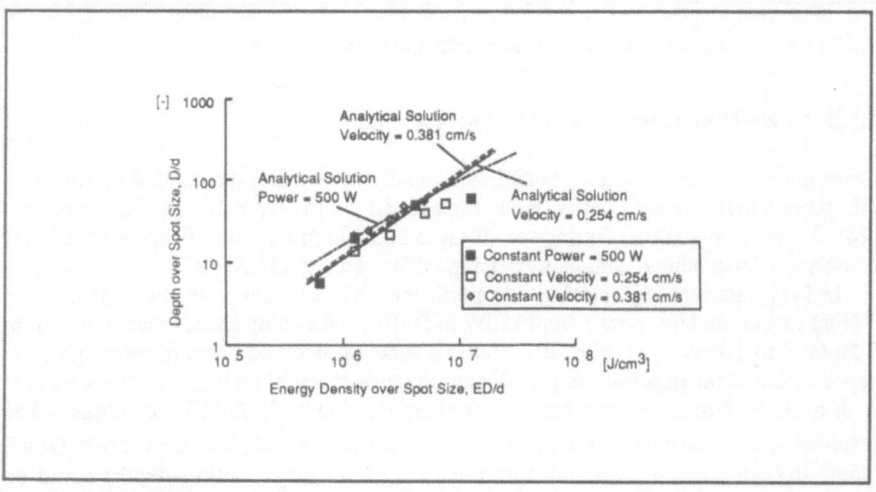

FIGURE 6.22   Cutting Depth vs. Energy Density for Alumina [8]

Material	Laser Type	Laser Power/ Power Density	Cutting Speed	Depth of Cut	Kerf Width	Refs
Aluminum Oxide	CO_2	150W	0.5m/min	0.635mm		41
Aluminum Oxide	Nd:YAG	$50MW/cm^2$	20mm/min	15mm		1
Aluminum Oxide	CO_2	$10MW/cm^2$	5mm/s	0.63mm		24
PSZ	CO_2	2-6kW	360-463 in/min	0.004-0.01in	0.036-0.1in	21
Quartz	CO_2	500W	1.2m/min	2mm		41
Silica	CO_2	450W	0.6m/min	1mm		41
Silicon Carbide	Nd:YAG	$6MW/cm^2$	40mm/min	5mm		1
Silicon Nitride	CO_2	1-2kW	60-123 in/min	0.02in	0.084in	21
Silicon Nitride	Nd:YAG	$10MW/cm^2$	50mm/min	3mm		1
Tungsten Carbide	Nd:YAG	$10MW/cm^2$	45mm/min	5mm		1
Zirconium Oxide	Nd:YAG	$5MW/cm^2$	200mm/min	1.3mm		1

TABLE 6.6   Representative Data for Laser Cutting of Ceramics

6.3.3 Two-Dimensional Machining: Scribing and Marking

Laser scribing has been used for high-speed marking of identification labels on ceramic parts (Table 6.7). Laser scribing has also been used in the electronics industry for producing snap lines in aluminum oxide wafers during fabrication of circuit boards [45]. Snap lines from 0.08 to 0.2 mm deep were formed using 50W and 125W pulsed CO_2 lasers with a focussed beam diameter of 140μm. The pulse frequency was 250Hz, pulse durations varied from 0.28 to 0.8 ms, and the scanning velocity was 80mm/s. Microscopic analysis of the workpiece surface showed the appearance of micro-cracks along the snap line. A thin layer of molten material was also found. Solidification of molten material led to material shrinkage, which resulted in residual stresses near the snap line. In [1], a scribe line 40μm deep was produced busing a 80W pulsed CO_2 laser and a scanning velocity of 5m/min. Micro-crack formation in the workpiece was minimized by reducing the pulse duration to 50ms. At this pulse duration, the boundary between the recast layer and the workpiece material served as an effective barrier to micro-crack propagation. In [24], 0.63mm deep snap lines were produced using both CO_2 lasers, with scribing speeds ranging from 15mm/s to 75mm/s, and Nd:YAG lasers, with scribing speed of 7.5mm/s.

In [6, 8, 9], grooves were made in aluminum oxide using a 1.2kW CO_2 laser operating in continuous wave. Spot diameters of 0.1cm and 0.013cm were used. Groove depths between 0.1mm and 20mm were achieved using energy densities between $10^3 J/cm^2$ and $5 \times 10^5 J/cm^2$. The energy efficiency, calculated as material's heat of fusion over laser input energy, was observed relative to the groove depth. The energy efficiency was found to be close to maximum up to a groove depth equal to four times the spot diameter (Fig. 6.23). For larger groove depths, the energy efficiency decreased For groove depths over 30 times the spot diameter, the energy efficiency was reduced considerably.

For deep grooves, formation of resolidified material along the groove walls becomes a significant effect; the resolidified material reduces energy efficiency by absorbing beam energy which would otherwise remove additional material. The recast layer, which also degrades the surface quality (Figs. 6.24 and 6.25), can be minimized by using an off-axial gas jet to expel molten material (Figs. 6.26 and 6.27). A series of grooves can be produced in an alumina workpiece to form cooling channels (Fig. 6.28).

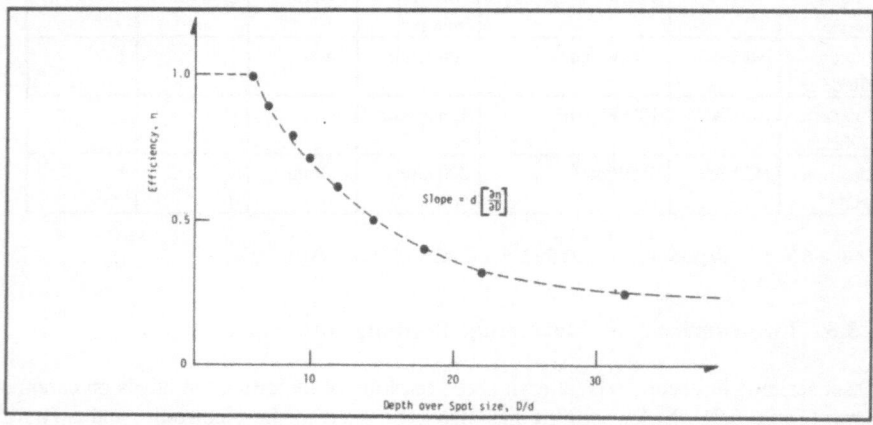

FIGURE 6.23   Energy Efficiency vs. Groove Depth for Alumina [6]

FIGURE 6.24   SEM Cross Section of Laser-Cut Groove in Alumina (30x)

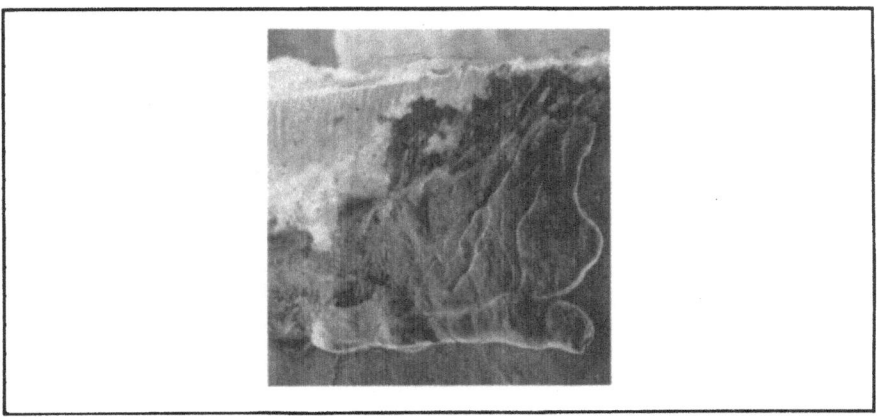

FIGURE 6.25   SEM Side View of Laser-Cut Groove in Alumina (190x)

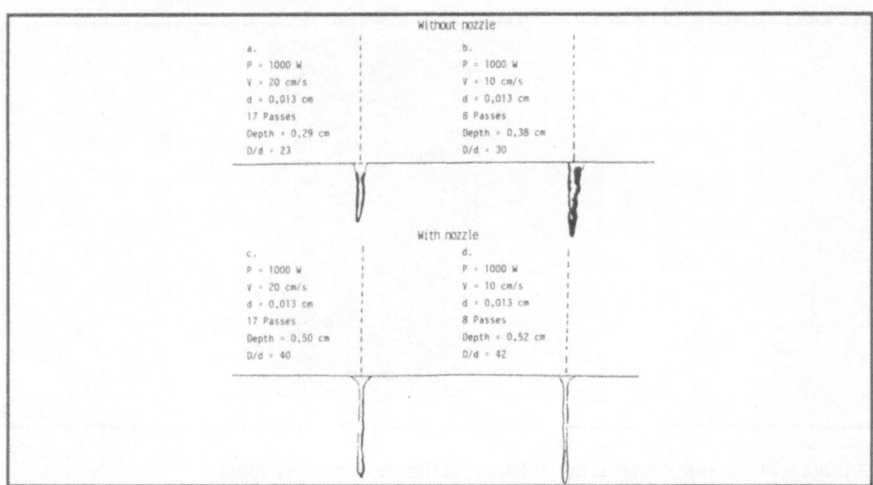

FIGURE 6.26   Cross-Sectional Groove Shapes for Various Gas Jet Conditions for Alumina [14]

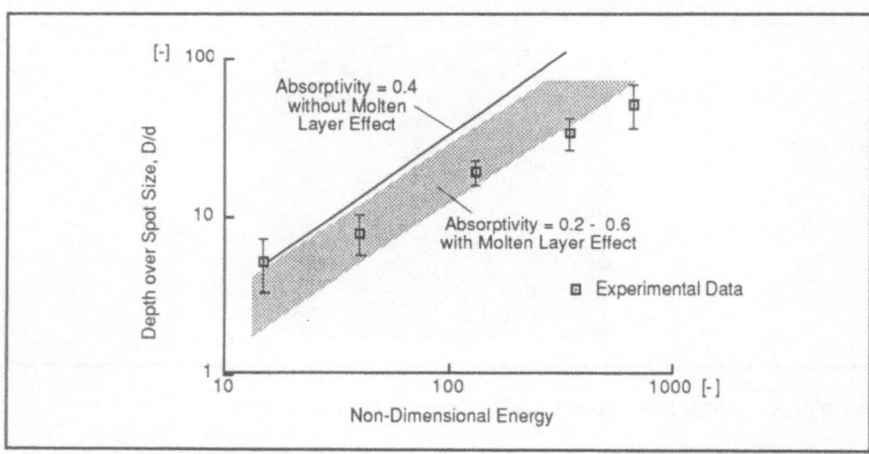

FIGURE 6.27   Groove Depth vs. Energy Density for Alumina [6]

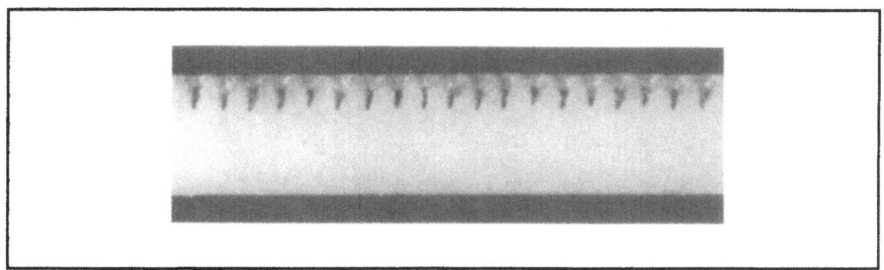

FIGURE 6.28   Laser-Cut Grooves in Alumina for Cooling Applications [14]

Laser scribing of silicon nitride workpieces was performed in [16] using a 1.2kW CW CO_2 laser. Scanning velocity ranged from 5cm/s to 125cm/s. Groove depths of up to 1mm were machined. Because of the use of a linearly-polarized beam, groove shape and depth were affected by the scanning direction of the beam. The deepest grooves were machined when the scanning direction was parallel to the direction of polarization. The minimum groove depth for a given operating condition occurred when the scanning direction was perpendicular to the beam polarization direction. For other scanning directions, the groove shape was shown to skew towards the polarization direction.

Tungsten carbide material can be scribed using high laser power and low scanning velocity. In [1], a 10kW Nd:YAG laser was used with a 30Hz pulse rate and a 0.1ms pulse duration. A 40μm deep scribe line was machined with a scanning velocity of 54mm/min. When the scanning velocity was reduced to 27mm/min, the width of the scribe line increased, while the distance between perforations decreased. The thermal damage to the workpiece could be reduced by using two beam passes at 54mm/min. Laser grooving has also been performed on zirconia (Fig. 6.29).

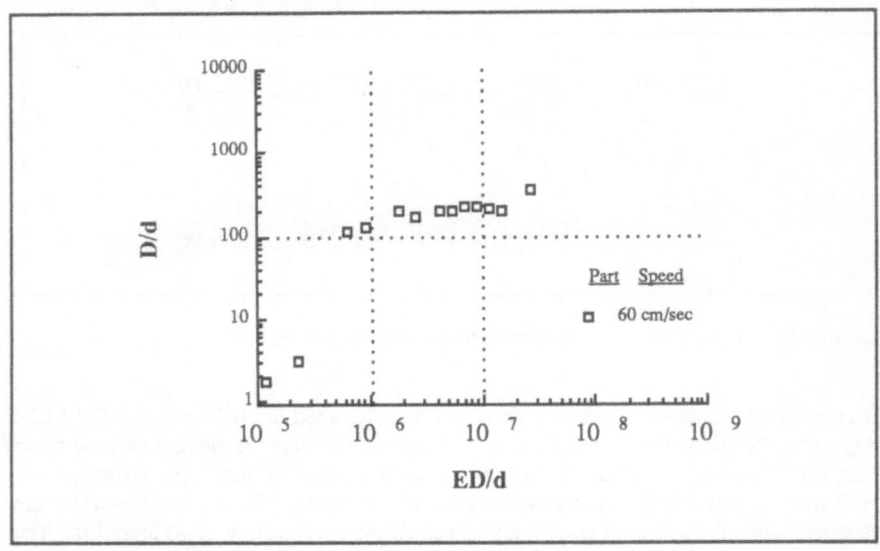

FIGURE 6.29   Groove Depth vs. Energy Density for Zirconia

An alternative method for laser scribing for producing snap lines is proposed in [46] and [10]. Using a CW CO_2 laser, the laser beam was used to either heat or melt a line on the surface of the workpiece. The rapid heating causes micro-crack formation along the line, and the workpiece can be separated by applying a bending torque. One disadvantage of this process is that impurities in the material will cause propagation of cracks outside the line, which will affect the straightness of the snapline. The advantage is that high scanning velocities can be used. For alumina substrates, scanning velocities of up to 5mm/s can be achieved with beam power densities between 10^4W/cm^2 and 10^5W/cm^2.

Material	Laser Type	Laser Power	Scanning Speed	Kerf Depth	Applica-tion	Refs
Aluminum Oxide	CO_2	50-125W	80-100 mm/s	0.04-0.2 mm	Snap Lines	1, 45
Aluminum Oxide	CO_2	1kW	0.5cm/s	5mm	17 passes	6
Aluminum Oxide	CO_2	1.2kW	0.05-5cm/s	0.1-20mm		6, 8, 24
Aluminum Oxide	Nd:YAG	10MW/cm^2	7.5mm/s	1mm		24
Silicon Nitride	CO_2	340-560W	5-38cm/s	0.01-0.5 mm		15
Tungsten Carbide	Nd:YAG	1kW	54mm/min	0.04mm		1

TABLE 6.7   Representative Data for Laser Scribing of Ceramics

6.3.4 Three-Dimensional Machining

A new approach has recently been developed for three-dimensional laser machining, as described in Section 6.2.4. The two-beam laser machining concept can be used to perform turning and milling operations on difficult-to-machine materials such as aluminum oxide and silicon nitride [6]. With this method only a small portion of the total volume is removed through melting; the majority of material is removed in the form of chips. Therefore, the energy efficiency of the process in terms of the volume of material removed per unit of energy expended is significantly better than that for other non-traditional machining processes. Also, this process produces three-dimensional geometries such as threaded or curved surfaces which are difficult to reproduce with mechanical processes. Due to the hardness and brittleness of most ceramics, mechanical machining processes usually result in a significant amount of tool wear, which increases the manufacturing cost of the part. Since the laser machining process does not involve contact between the tool and workpiece, there is no tool wear or tool breakage. During the laser machining process, molten material is ejected from the grooves with the aid of supersonic off-axial gas jets. Similar to three-dimensional laser machining of metals, high energy densities and multiple beam passes are required for chip removal in ceramics.

A single-beam pocket cutting technique can be used to remove a volume of material through selective melting or vaporization with a laser beam. In [1], a 0.7mm deep pocket was machined in silicon nitride by machining overlapping grooves with the laser beam perpendicular to the workpiece surface. A power density of $6 \times 10^6 W/cm^2$ was used, resulting in a material removal rate of $30mm^3/min$; a surface finish of $7\mu m$ RMS was achieved.

In [16], a single-beam technique for creating three-dimensional contours is presented; a 450W CW CO_2 laser was used to machine overlapping grooves on a ceramic workpiece in order to remove a volume of material. Silicon carbide, silicon nitride, alumina, and SiAlON were tested. The beam was directed tangentially to the workpiece surface in order to create a groove. In the study, the surface roughness of the finished part was controlled by decreasing the groove depth on successive overlapping passes. This technique was used to produce flat or threaded surfaces on a workpiece. The process described in [16] is analogous to electro-discharge machining (EDM) processes, since beam energy must be used to melt or vaporize the entire material volume to be removed. Material removal rates can be increased by using a N_2 or O_2 gas jet for reactive gas cutting.

6.4 Plastics

6.4.1 One-Dimensional Machining: Drilling

Lasers can be used for many deep or high-precision hole drilling applications in plastics, such as drilling aerosol nozzles or catheters. Because most plastics have high absorptivity values at the wavelengths of CO_2 and Nd:YAG laser radiation, the energy losses due to surface reflection are low. Since material is primarily removed through vaporization, the problems of molten material accumulation and resolidification associated with drilling of metals and ceramics are not present in plastics. Also, the threshold energy density for material removal in plastics is several orders of magnitude lower than the threshold for ceramics, so that relatively deep hole drilling can be performed with lasers (Table 6.8). The primary cocern in laser drilling of plastics is hole taper as well as thermal damage in the case of thermosetting plastics. Since deep holes can be easily drilled, one problem with laser drilling is the taper associated with the hole due to the divergence of the laser beam. In applications where straight-sided holes are essential, the hole taper can be minimized if beam stability is controlled by maintaining a stable TEM_{00} mode and by continuously shifting the focal point beneath the top surface of the workpiece. Thermal degradation occurs when the temperature of the workpiece is raised above the charring temperature, where the polymer material decomposes due to chemical interactions such as oxidation or pyrolysis into gaseous and char components.

For aerosol nozzles, holes with diameters from 0.15mm to 1mm are required. Conventional molding techniques do not have the flexibility to change hole size in process. Laser drilling combines the flexibility of changeable hole diameter with high production rates [14]. Using a CO_2 laser with power ranging from 50W to 185W, drilling at rates between 20 to 1000 holes per second can be achieved. Hole diameter variations are less than ±0.025mm from part to part. The hole diameter can be changed in process by either changing the focal distance between the laser beam and the workpiece or changing the beam power.

Material	Laser Type	Laser Power	Drilling Rate	Hole Depth	Hole Diameter	Application	Refs
Mylar	CO_2	185W	0.15ms/ hole	0.01in	0.01in		14
Nylon	CO_2	185W	0.5s/hole	0.3in	0.008in		14
Poly-ethylene	CO_2	50-185W	0.05-8ms per hole		0.006-0.1 in	Flow Meter/ Aerosol Nozzle	14
Poly-urethane	CO_2	185W	0.6s/hole	0.015in	0.1in	Catheter Drilliing	14

TABLE 6.8   Representative Data for Laser Drilling of Plastics

6.4.2   Two-Dimensional Machining: Cutting

Lasers can be used for high-speed cutting of plastic sheets (Table 6.9). The primary factors in laser cutting of plastics are melt shearing (flow of molten material out of the kerf under the influence of a gas jet), vaporization, and chemical degradation [41]. For the low energy densities investigated in [41], melt shearing was observed to be the primary mechanism for material removal for many common polymers, including polypropylene, polystyrene, polyethylene, nylon, and ABS. Melt shearing is similar to the cutting mechanism for ceramics and metals, where the laser beam melts the material at the cutting front and the molten material is ejected with a powerful gas jet. Vaporization occurs during laser cutting of materials with high absorptivity such as acrylic (Fig. 6.30) or when high energy densities above $10^5 J/cm^2$ are used. Chemical degradation occurs mainly when laser cutting is applied to epoxy or phenolic resins. The feasibility of laser cutting for a wide range of thermoplastic and thermoset materials was examined in [52, 53] for variations in power, cutting speed, material thickness, kerf width at the kerf bottom, edge taper (difference between the top and bottom kerf widths), and charring. In general, cutting speeds up to 3000mm/min can be achieved for most plastics. Beam power can vary from 50W to 950W.

Laser cutting of thermoset materials produces a char layer along the kerf, usually less than 0.05mm in thickness. The charred layer is most evident in laser cutting of epoxy, polyester, phenolic, and polycarbonate materials. The char layer is removed in a secondary operation, either with an alcohol wipe or glass bead vapor blast. Thermoplastic materials can generally be processed without charring.

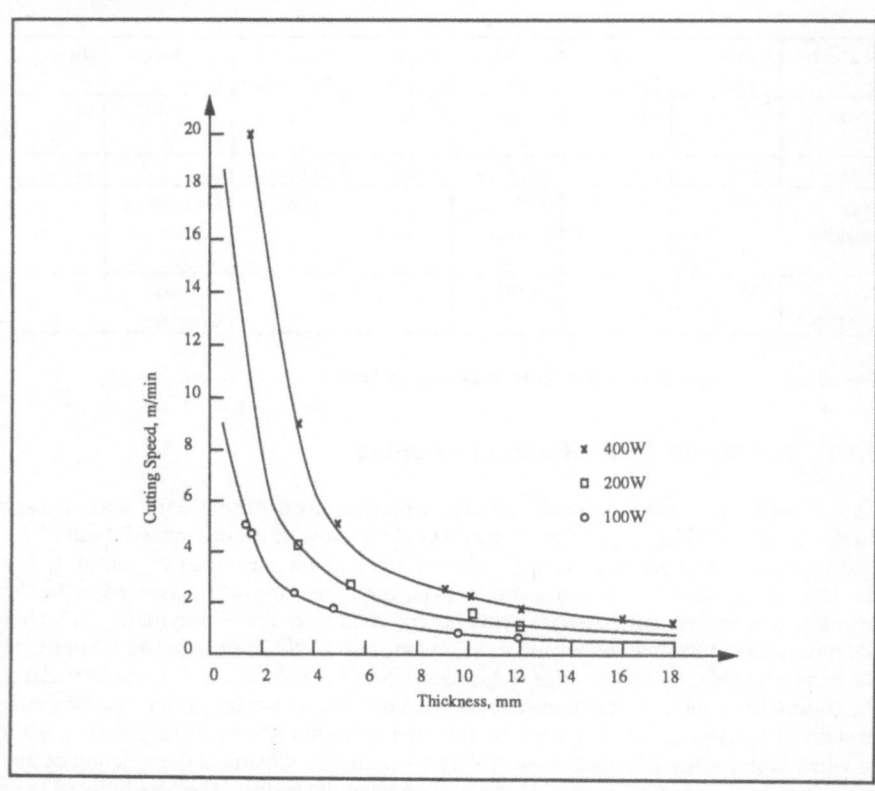

FIGURE 6.30   Cutting Speed vs. Workpiece Thickness for Acrylic

One particularly successful industrial application for plastics is laser cutting of seat belts [14], which are usually manufactured from woven nylon, which is difficult to cut mechanically. Using a 500W beam, cutting speeds of over 12m/min were achievable. The machined edge is sealed with no burring or charring. Another industrial application is the selective removal of plastic coating from a metallic film or wire. The shield material can be vaporized with no damage to the metal underneath.

Material	Laser Type	Laser Power	Scan. Speed	Kerf Depth	Refs
ABS	CO_2	400-1200 W	0.6-4.5 m/min	3.175-12.7 mm	5.41
Poly-carbonate	CO_2	400W	0.82-4.8 m/min	3.175-12.7 mm	5.41
Poly-ethylene	CO_2	400W	0.4-18 m/min	0.635-12.7 mm	5.41
Poly-propylene	CO_2	400W	0.5-3.4 m/min	3.175-12.7 mm	5.41
Poly-styrene	CO_2	400-1300 W	0.55-50 m/min	0.635-7.5 mm	5.41

TABLE 6.9   Representative Data for Laser Cutting of Plastics

6.4.3   Two-Dimensional Machining: Scribing and Marking

Laser scribing and marking of plastics has been used for production of templates and marking of identification labels on plastic parts (Table 6.10). Laser scribing processes for engraving identification labels on plastic packages were investigated in [58]. A Nd:YAG laser was used with energy density set between 10^4 and 10^7 W/cm^2. Material removal was performed through vaporization. At high energy densities near 10^7 W/cm^2, some material removal also occurred through small explosions at the erosion front. Pulse rates were set for 70% to 90% pulse overlap. Scribing depths ranged from 1mm to 10mm. In many cases, compressed air is used as an assist gas to increase groove depth and increase charring of the material to make the engraving more legible. Results on laser scribing using a 1kW CO_2 laser were reported in [14].

One successful industrial application is the production of acrylic positioning templates for drafting. The curved guide lines on each template are scribed into a flat workpiece with a laser beam. Laser scribing produces clean guide lines with high dimensional accuracy and no debris.

Material	Laser Type	Laser Power/ Power Density	Groove Depth	Applica-tion	Refs
Acrylic	CO_2	1kW	1mm	Template	14
Poly-ethylene	Nd:YAG	$10kW/cm^2$- $10MW/cm^2$	1-10mm	Marking Packages	58

TABLE 6.10   Representative Data for Laser Scribing of Plastics

6.4.4   Three-Dimensional Machining

The three-dimensional laser machining concept described in [7] can also be applied to shaping of plastic parts. The three-dimensional process presents a method for creating complicated three-dimensional shapes in plastics with high machining speeds. Since most plastic materials have high absorptivity for CO_2 and Nd:YAG laser wavelength irradiation, deep grooves can be produced and large material volumes can be removed with only one beam pass, compared with the multiple beam passes required to remove a volume of material for metals and ceramics.

Helical removal, ring removal, and translational milling techniques can be used to process acrylic parts (Fig. 6.16). Threads with 60° angle and 3.2mm pitch were produced on cylindrical acrylic workpieces. For rectangular workpieces, laser milling using two laser beams was implemented (Fig. 6.31). A comparison of material removal rates versus surface quality shows a tradeoff between surface roughness and material removal rate (Fig. 6.32).

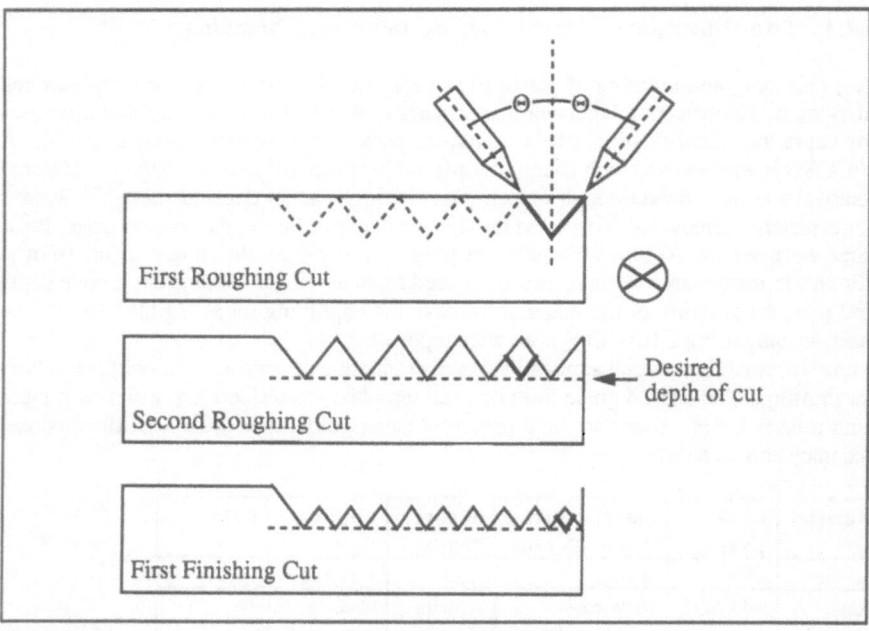

FIGURE 6.31   Configuration for Two-Beam Milling of Acrylic

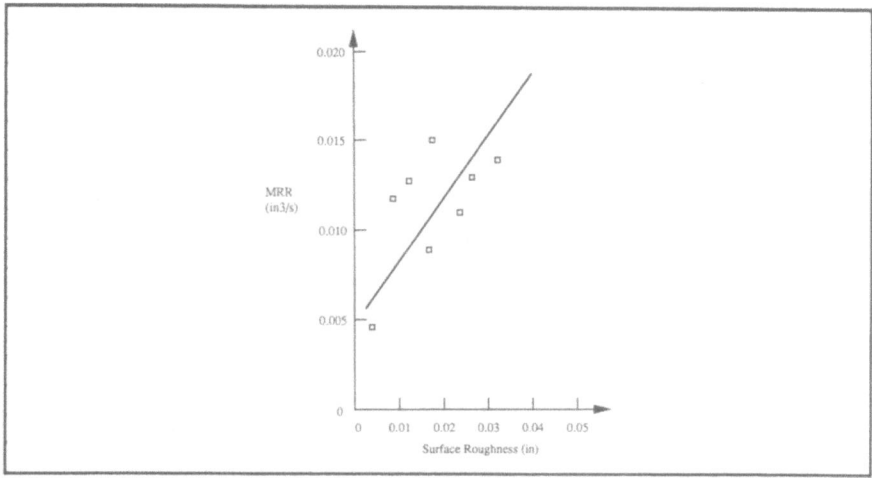

FIGURE 6.32   Material Removal Rate vs. Surface Roughness for Multiple-Pass Laser Milling of Acrylic

6.5   Composite Materials

6.5.1   One-Dimensional Machining: Drilling

Composites have a great potential for use in manufacturing applications due to their light weight, high strength, and directional properties. Although in most applications, composites are cured in a mold to a "near final shape," machining is still required at both the prepreg and finishing stages. A finishing process is usually necessary to achieve the final surface quality and dimensional accuracy if the required tolerances are high and cannot be achieved through the initial processing of the composite part. Finishing or secondary processes usually include deburring, trimming, drilling and boring. For deburring and trimming a typical accuracy to be achieved is ±1mm; while for drilling and boring operations, accuracies of less than ±0.1mm are required.

Conventional machining operations are difficult to perform on composite materials due to their anisotropy, inhomogeneous composition, hardness, and abrasiveness. Delamination, debonding, and fiber breakage can lead to strength reduction for the finished part. Excessive tool wear significantly increases the machining time and expense. Laser machining offers the advantages of high machining rates, no tool wear, no contact forces, and relative high precision. A comparison of various techniques for performing secondary operations on composites is shown in Figure 6.34.   The effectiveness of laser machining depends primarily on the thermal properties of the

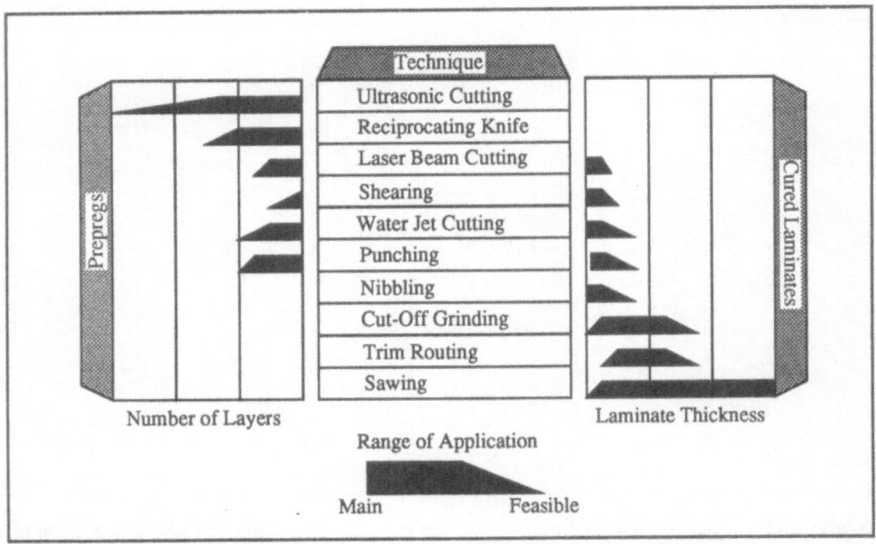

FIGURE 6.33   Comparison of Machining Techniques for Composite Materials [28]

material; the type and amount of constituent fiber and matrix materials determine the thermal response of the composite material to the laser beam.

Laser drilling has been applied to many Kevlar and graphite-fiber composite materials (Table 6.11). For most composites, mechanical machining techniques create excessive heat buildup and shear forces at the cutting zone, causing delamination between composite layers and debonding between fiber and matrix. In [33], 0.125in diameter holes were drilled in 0.050in thick Kevlar/epoxy for a 1200W beam using the trepanning method. Cycle time was approximately 1.5 sec per hole. In [14], a similar method was used with a 185W pulsed beam to produce 0.2in diameter holes in 0.0in thick Kevlar/epoxy. Cycle times of 0.6 sec/hole was achieved. Laser drilling was also conducted on several ceramic matrix composites. Silicon nitride and alumina matrix materials with imbedded zirconia were drilled with a 1200W beam to produce mounting holes for the manufacture of tool inserts.

Material	Laser Type	Laser Power	Drilling Rate	Hole Depth	Hole Diameter	Application	Refs
Kevlar/ Epoxy	CO_2	185W-1.2kW	0.6-1.5 s/hole	0.05in	0.125-0.2 in	Trepanning	14, 33

TABLE 6.11   Representative Data for Laser Drilling of Composites

6.5.2   Two-Dimensional Machining: Cutting

Laser machining often competes with water jet and mechanical cutting techniques for finish machining applications. Although laser systems usually have a high capital cost, laser cutting has the advantages of a narrow kerf width, localized material damage, and high cutting speeds. For aramid-fiber materials, laser machining rates of 2.5 times mechanical cutting speeds can be achieved [28].

Several recent investigations have been performed on laser cutting and drilling operations for fiber-reinforced polymers (Table 6.12). Cutting on glass/polyester, graphite/polyester and aramid/polyester materials [48, 49, 22, 28] was performed for material thicknesses ranging from 2.0mm for graphite/polyester to 4.5mm for Kevlar/polyester. Laser power ranged from 300W to 2kW. A coaxial Helium gas jet was used. Laser scanning velocities varied from 0.5 to 3.5m/min. Using a 1kW CW beam and scanning velocities in the range of 17 to 34mm/s, through-cutting was performed on Kevlar/epoxy workpieces with thicknesses up to 9.5mm [51]. The laser power required to cut multi-layered Kevlar/epoxy laminate can be seen in Figure 6.35. The kerf width for different workpiece thicknesses is shown in Figure 6.36. The use of a pulsed beam significantly improved the cutting speed for glass-fiber materials [28]. One characteristic of laser cutting is the relation between cutting efficiency and beam scanning direction relative to fiber orientation. Due to material anisotropy, the thermal response of a workpiece to the laser beam depends on cutting direction. This effect is most apparent in graphite-fiber composites, where the thermal properties between the constituent materials are very different. In cutting experiments on a uniaxial laminate [22], heat losses were found to be maximum and cutting speed were minimal when the fibers were orthogonal to the cutting direction; in this orientation, the heat flux is along the fiber direction.

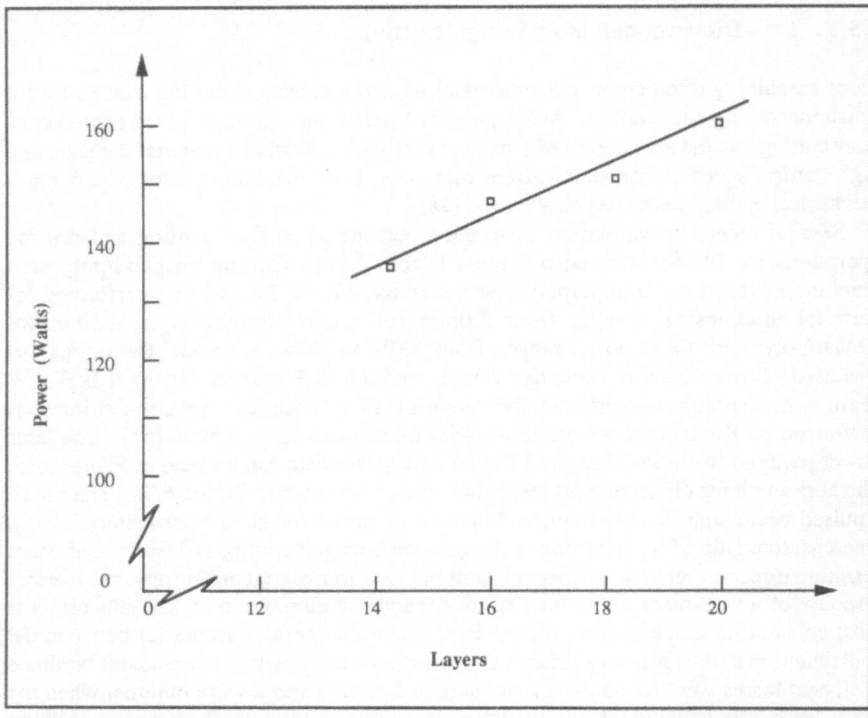

FIGURE 6.34   Laser Power vs. Number of Layers for 3.1mm Thick Kevlar/Epoxy [51]

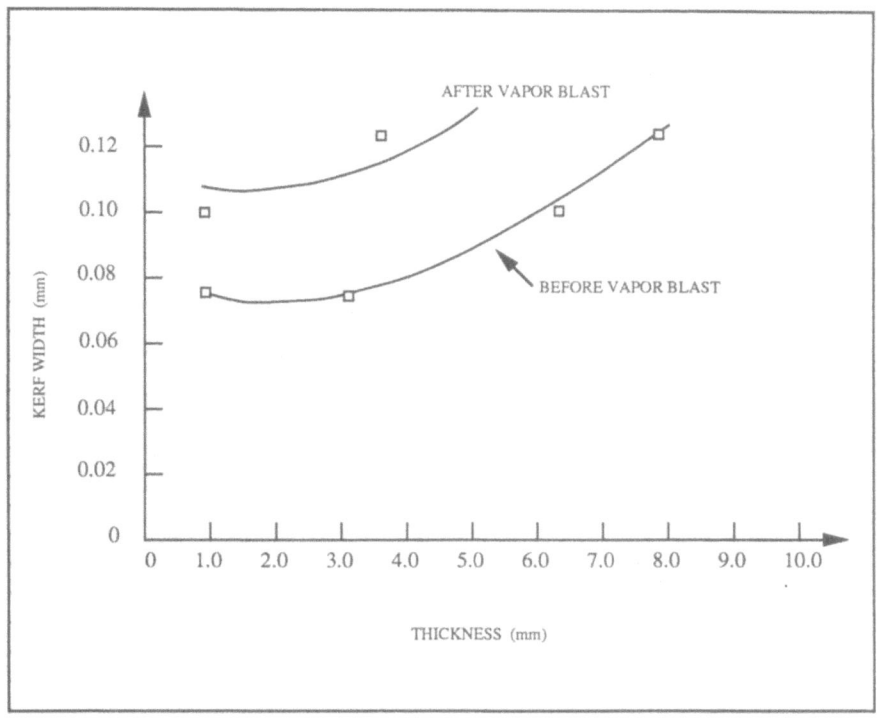

FIGURE 6.35   Kerf Width vs. Thickness for Laser-Cut Kevlar/Epoxy [51]

Issues regarding kerf quality for aramid, graphite, and hybrid aramid/graphite reinforced polymer materials were investigated in [28] and [48, 49]. An international cooperative effort on determining cutting quality led to the selection of parameters shown in Figure 6.37 as important for evaluating the quality of composite parts cut by a laser beam. Cutting quality, in terms of presence of charred material, extension of delamination, and slope of the kerf surface, depend on the scanning velocity, laser power, and the assist gas. A coaxial inert gas jet (He and Ar) can be used in the cutting process to minimize charring. In general, the cutting quality is highest for composites which exhibited similar thermal properties between constituent materials (such as Kevlar/polyester). Graphite and glass-fiber materials yield poorer results due to the high fiber vaporization temperatures and the high thermal conductivity of graphite. Graphite-reinforced composites are found to be less suitable for laser cutting due to high fiber conductivity and vaporization temperature (similar to the results from [49]). Two modes of damage were found for the cutting process. Surface damage due to heat conduction effects (primary damage mode) was found to exist near the beam entrance region of the kerf. Surface damage due to reflected assist gas (secondary damage mode) occurred mainly near the beam exit region. In laser cutting experiments performed on graphite/epoxy, the heat-affected zone at the kerf entrance was found to be significantly higher than at the kerf exit (Fig. 6.37) [28]. The thermal damage zone for glass and aramid-fiber composites was only 20% to 30% as large as the damage zone for graphite-fiber composites (Fig. 6.38); the thermal damage width also showed a decrease with increase in cutting speed [49].

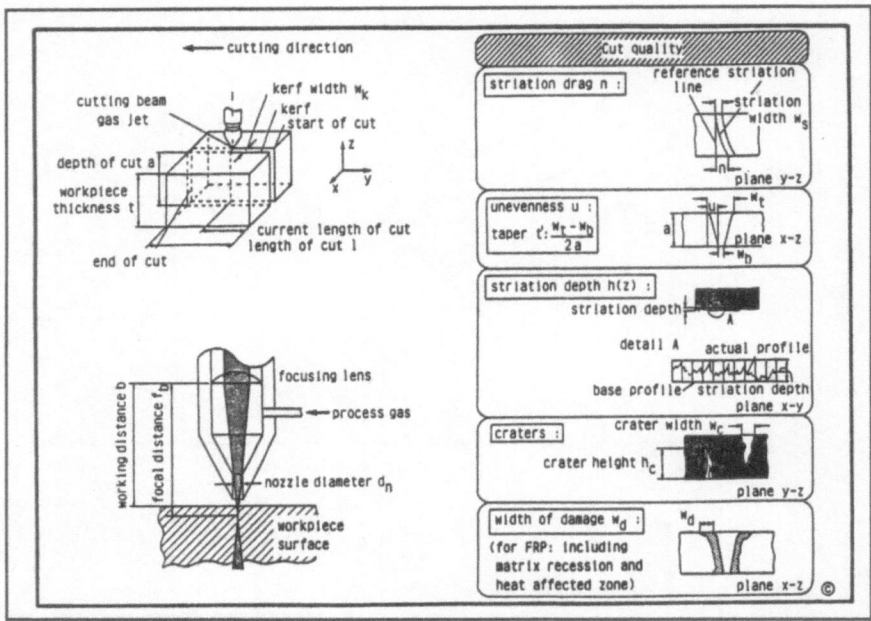

FIGURE 6.36   CIRP Standards for Cutting Quality for Laser Machining of Composite Materials [48]

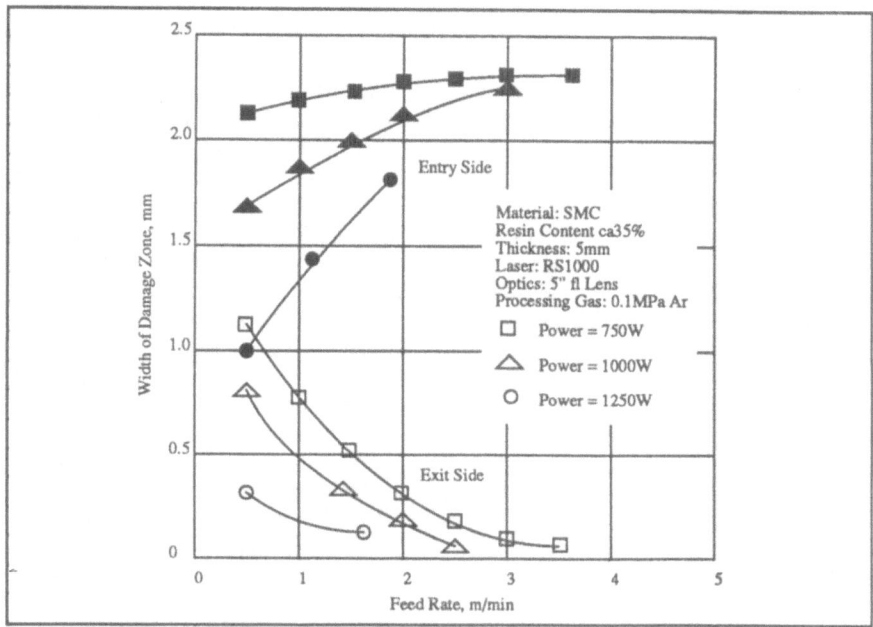

FIGURE 6.37   Damage Width vs. Feed Rate for Laser Cutting of Graphite/Epoxy [28]

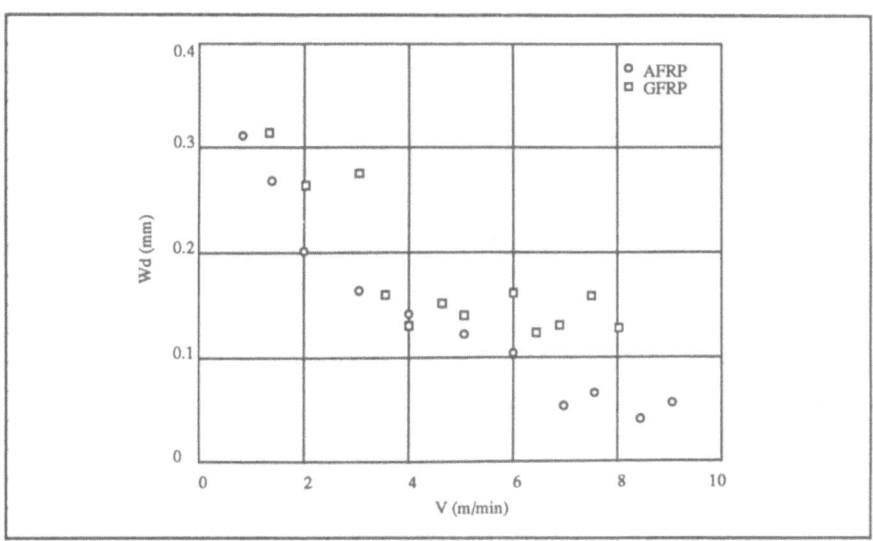

FIGURE 6.38   Damage Width vs. Feed Rate for Laser Cutting of Glass/Polyester and Aramid/Polyester [48]

Some common characteristics of surface morphology encountered in experimental studies are matrix recession, namely uneven material loss between fiber and matrix when heated by a laser beam, formation of a carbonized layer near the kerf surface, and nonuniform lengths for the fibers on the kerf.   In most cases, surface charring contributes to an increase in edge taper and surface roughness.   In [51] mechanical abrasion post-processing was used in order to reduce the carbonized layer; the use of high scanning velocities to reduce thermal conduction damage is suggested in [28].   Nonuniform material loss appeared to be a function of the matrix and fiber materials selected (aramid versus glass and graphite materials) and depends strongly on the thermal conductivity of the constituent materials.   Aramid fibers have thermal characteristics similar to polymer matrix materials, so material behavior during beam/material interaction is similar to that of homogeneous materials.   SEM micrographs of the kerf surface for aramid/epoxy shows a relatively smooth surface between fiber and matrix regions (Fig. 6.39).   Graphite fibers, however, exhibit thermal conductivity values which are up to two orders of magnitude higher than polymer matrix materials.   This leads to highly directional heat conduction in the workpiece and a much larger thermal damage zone than with other fiber materials.   Also, the thermal damage zone tends to propagate along the direction of fiber orientation.   Glass fibers can be removed through melting, instead of vaporization.   This phenomenon can lead to increased surface roughness, or irregularities in fiber tip lengths, due to the presence of resolidified molten material.

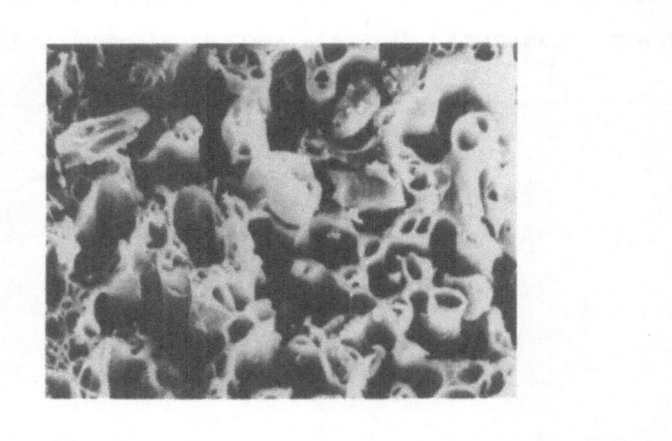

FIGURE 6.39   SEM of Laser Cut Surface for Aramid/Epoxy [22]

Material	Laser Type	Laser Power	Cutting Speed	Kerf Depth	Kerf Width	Refs
Aramid/ Polyester	CO_2	800W	0.5m/min	2mm	0.6mm	49
Glass/ Epoxy	CO_2	1kW	2m/min	5mm	0.5mm	28
Glass/ Polyester	CO_2	800W	0.5m/min	2mm		49
Graphite/ Epoxy	CO_2	1-2kW	15-120 mm/s	1-4mm		22
Graphite/ Epoxy	CO_2	300W	5mm/s	1mm	0.1mm	22
Graphite/ Polyester	CO_2	800W	0.5m/min	2mm	0.5mm	49
Kevlar/ Epoxy	CO_2	150-950W	34mm/s	3.2-9mm	0.1mm	51

TABLE 6.12   Representative Data for Laser Cutting of Composites

One problem in laser machining of composite materials is that the products of chemical decomposition produced under elevated temperatures can be hazardous. Mass spectrometry and gas chromatography analysis were conducted on gas samples during laser cutting of graphite/epoxy, aramid/epoxy and glass/epoxy [22]. Most particulate products were fragmented powders of fiber materials. Matrix decomposition yielded concentrations of CO, CO_2 and low molecular organic compounds. It has been shown that cutting of aramid/epoxy produces large quantities of hydrogen cyanide, which may pose a considerable health risk.

6.5.3   Two-Dimensional Machining: Scribing

In several publications, the laser grooving or scribing process for carbon/teflon, glass/teflon and glass/polyester materials were investigated (Table 6.13) [11, 12, 13]. The teflon matrix materials (Fig. 6.40) were composed of short fibers in random orientation, while glass/polyester (Fig. 6.41) was composed of unidirectional long glass fibers. The major issue in laser scribing of composites is the anisotropic nature of the material. The groove depth, groove shape, and heat-affected zone will vary depending on the laser beam direction relative to fiber orientation. This effect is apparent for long-fiber composites where the fibers, with different thermal properties than the matrix, are oriented in specific directions within the matrix. In most grooving applications, the laser beam is directed perpendicular to fiber orientation since fibers are usually aligned in the plane perpendicular to the workpiece thickness; however, in some cases such as laser turning or threading, the laser beam can be directed at an arbitrary angle relative to fiber orientation. Heat losses to the environment and matrix decomposition have the effect of groove depth reductionparticularly at high energy densities (Figs. 6.40 and 6.41).

Material	Laser Type	Laser Power/ Power Density	Groove Depth	References
Carbon/ Teflon	CO_2	$100kW/cm^3$- $1MW/cm^3$	0.6-7mm	12
Glass/ Teflon	CO_2	$100kW/cm^3$- $1MW/cm^3$	0.9-7.1 mm	12
Glass/ Polyester	CO_2	$60kW/cm^3$- $1MW/cm^3$	0.62-4.5 mm	11
Graphite/ Vinylester	CO_2	$60kW/cm^3$- $1MW/cm^3$	0.53-6.2 mm	13

TABLE 6.13   Experimental Results for Laser Scribing of Composites

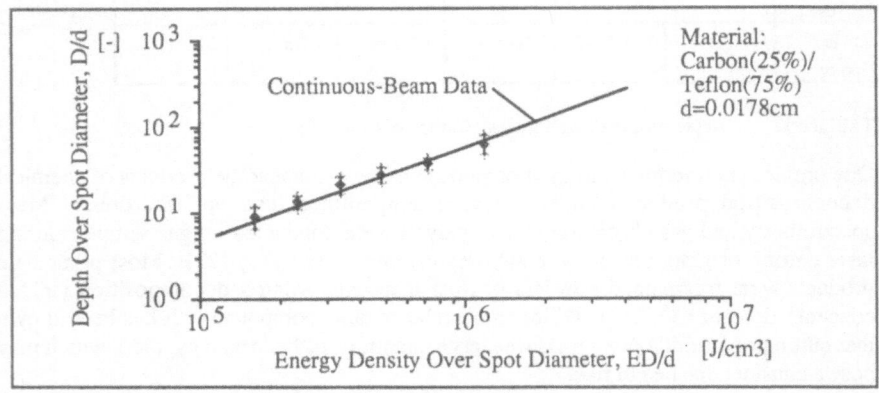

FIGURE 6.40   Groove Depth vs. Energy Density for Carbon/Teflon [12]

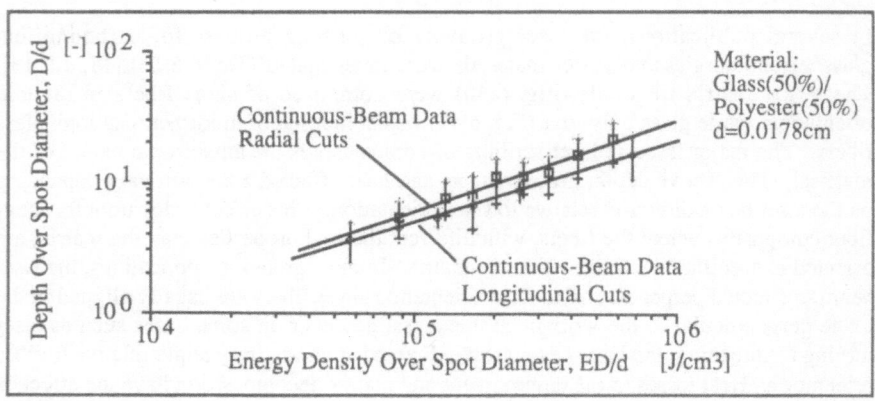

FIGURE 6.41   Groove Depth vs. Energy Density for Glass/Polyester [12]

6.5.4 Three-Dimensional Machining

Using the concept described in [13], threaded and turned workpieces can be produced for carbon/teflon and glass/teflon materials. A CW CO_2 laser beam operating at 300W, a feed rate of 0.1cm/s, and three beam passes was used to produce both threaded and turned workpieces with no visible charring or surface damage.

A single-beam pocket cutting method was introduced for machining kevlar/epoxy in [51]. In pocket cutting, a volume of material on a workpiece is removed by producing a set of parallel, overlapping grooves with a scanning laser beam. The groove centers are positioned 0.25mm apart. A pulsed 1kW CO_2 laser was used with 10ms pulse length and 67% duty cycle. A contoured cutout with a mean depth of 3.3mm could be produced in a 6.2mm thick laminate; the laser penetration depth increases with laser power increase (Fig. 6.42). The pocket depth can be controlled to ±0.25mm. The charred layer is removed through a secondary process using glass bead vapor blast.

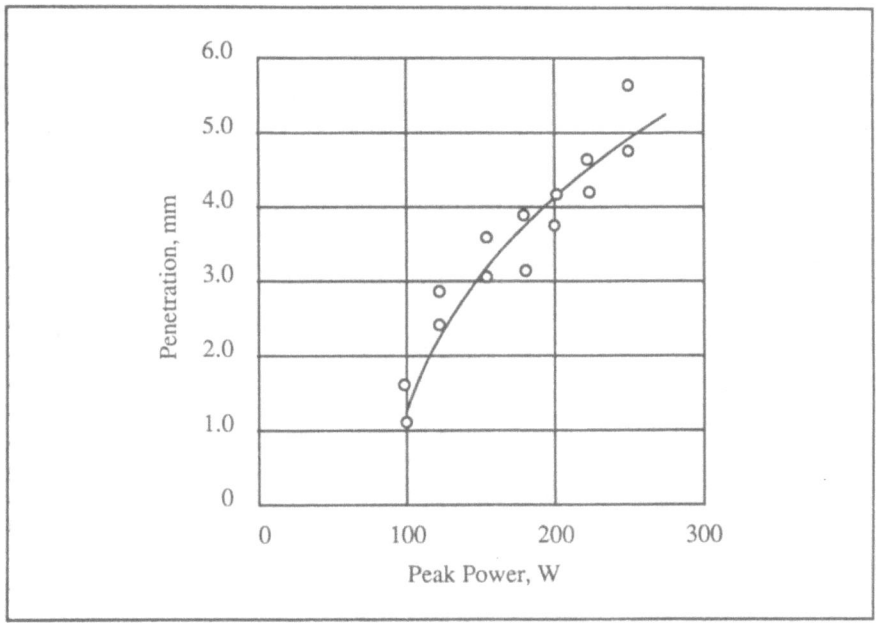

FIGURE 6.42   Laser Beam Penetration Depth vs. Power for Pocket Cutting [51]

6.6   Other Materials

Besides machining metals, ceramics, plastics, and composites, lasers have also been used for drilling, cutting, and scribing of wood, paper, glass, and rubber. The advantages of laser machining for these materials include high machining speeds, high dimensional accuracy, and no accumulation of debris. Some representative data is shown in Table 6.14.

6.6.1   Wood

Laser systems can be used for high-speed pattern cutting of furniture parts and dieboards. In most cases, laser cutting involves either vaporization or high-temperature oxidation or burning of the material. Laser cutting has several advantages over sawing, such as small kerf widths, the ability to cut complicated geometries and sharp corners, the ability to instantaneously start and stop cutting and to have a minimal effect of grain direction on cutting. Additionally, it eliminates dust, blade wear, the cutting force on wood and rough or torn-out grains. However, laser processing of wood can produce substantial charring near the kerf, which reduces the appearance of the finished part. This effect is usually reduced by using an inert gas jet with the laser beam. In general, values for the laser cutting parameters are governed by three parameters: workpiece thickness and the material's water and air content. Since water is highly absorptive to CO_2 laser radiation, a high water content decreases the cutting efficiency. The air content affects the density and thermal conductivity of the material.

Laser cutting of various types of wood with a 1kW CO_2 laser and a N_2 gas jet was investigated in [38] (Fig. 6.43). For a laser power of 250W, cutting speeds can vary according to the type of wood from 0.06m/min for red oak with 80% moisture content to 1m/min for soft maple [38]. Lower cutting speeds are generally required for wood with high moisture content due to the additional energy required to vaporize water. Knots in wood also required lower cutting speeds due to higher density values, but the resulting kerf shows no change in geometry from clear wood. Cutting speeds for particle board and plywood show a strong dependence on the type of binder used in the material. Materials with a urea binder can be cut up to three times faster than materials with a phenolic binder (Fig. 6.44). A thin char layer may also be formed along the kerf wall for particle board and plywood.

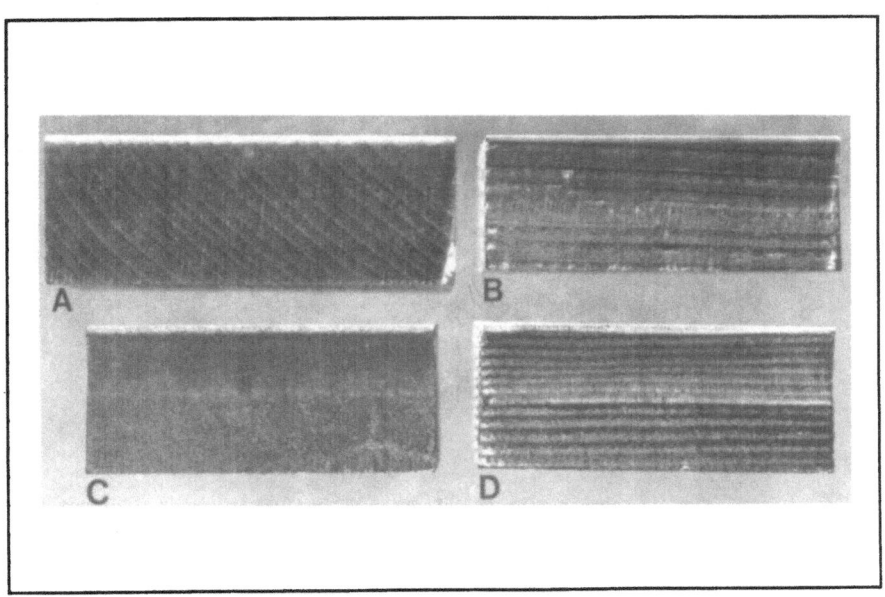

FIGURE 6.43   Laser-Cut Surfaces for 0.063in Thick (a) White Pine, (b) Southern Pine, (c) Hard
Maple, and (d) Douglas Fir [38]

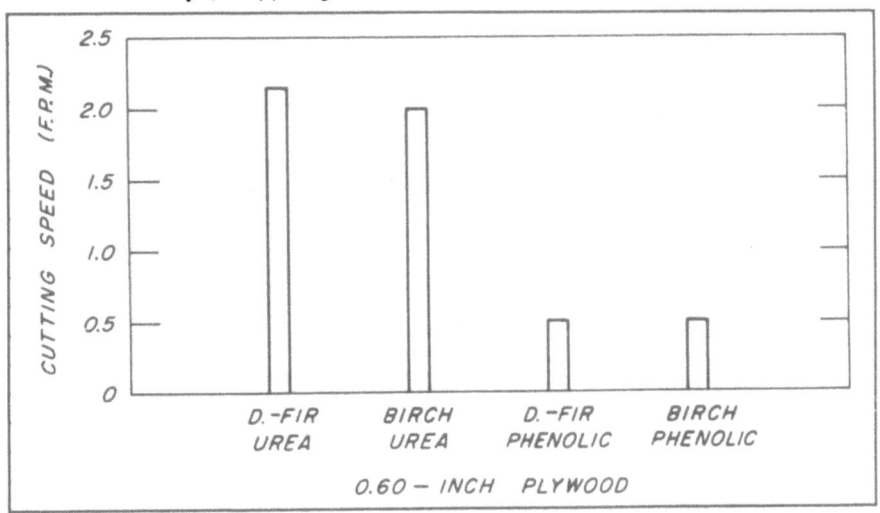

FIGURE 6.44   Laser Cutting Speeds for Plywood [38]

Lasers are used to manufacture dieboards for stamping paper cartons and boxes (Figs. 6.45 and 6.46). A 400W CO_2 laser beam can be used to cut slots in the shape of a die pattern in plywood. Metal blades are then inserted into the slots to produce the stamping dies.; accuracy of ±0.1mm can be achieved.

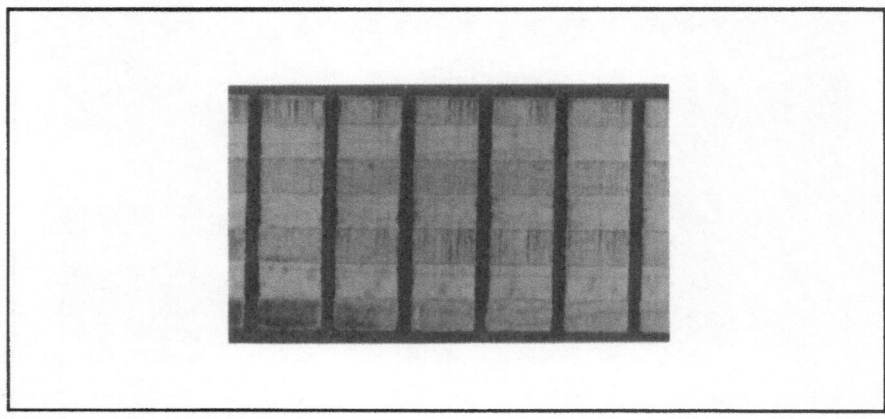

FIGURE 6.45   Laser-Cut Slots in Die-Board [14]

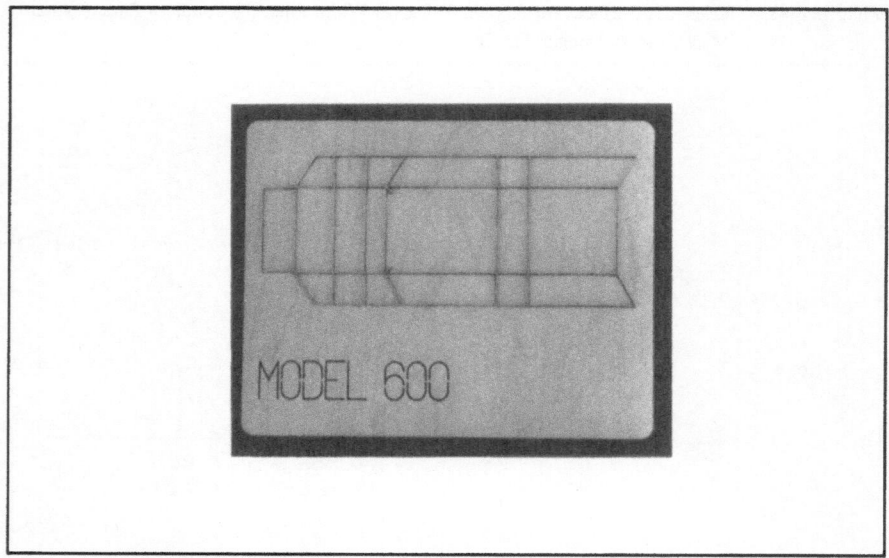

FIGURE 6.46   Die-Board Configuration [14]

6.6.2 Paper

The laser power and cutting speeds required for cutting paper can vary widely depending on the type of paper (Fig. 6.47). Mechanical cutting of paper usually results in accumulation of debris, which creates a significant health hazard and a waste disposal problem. In laser cutting the paper is vaporized, so no debris is left behind. Paper cutting and perforation with a laser has been used in the manufacture of cigarette paper. Hole drilling in cigarette paper requires a process that can create small-diameter holes precisely, since the hole size in cigarette paper controls the rate of passage of tar and nicotine into the filter and affects the tar and nicotine rating of the cigarette. Scanning velocities of 330m/min can be reached with a 185W beam [14]. Laser perforation can be performed by either using a pulsed beam or by placing a perforated metallic mask between the laser beam and paper (Fig. 6.48). Laser perforation produces much cleaner holes than mechanical perforation (Fig. 6.49). Laser cutting of other types of paper, such as newsprint and bond paper, may not be economical due to the high speeds achievable with mechanical cutters (up to 2300m/min per cutter with seven cutters per machine). Laser beam powers up to 70kW are required to produce equivalent cutting speeds [18].

Type of Paper	Thickness $(X\ 10^{-3}\ in.)$	Rate (FPM)
Bond	1.5	700
	2	1000
	3.5	200
	4	350
Graph paper	2	600
Newsprint	3	530
Coated paper	5	200
Forbes cover stock	4.5	300
Laminate	10	130
Bond (3 layers)	4.5	130

FIGURE 6.47   Maximum Cutting Speed for Paper With 250W CO_2 Laser [18]

FIGURE 6.48   Setup for Laser Perforation of Paper Using Metallic Mask [14]

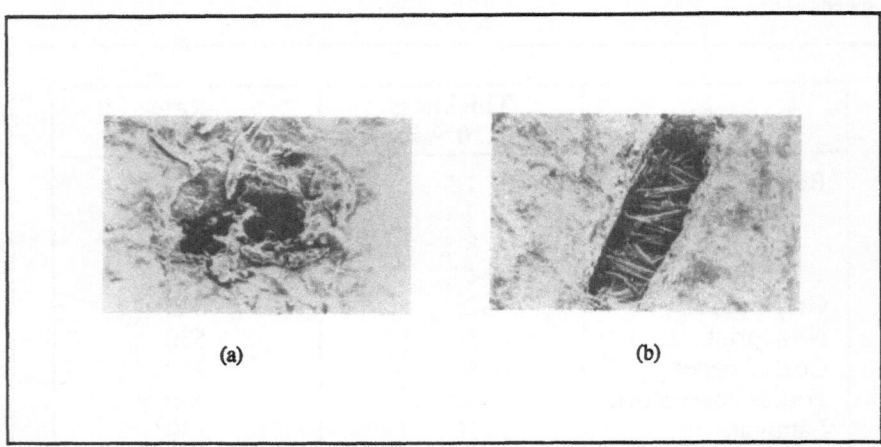

(a)   (b)

FIGURE 6.49   Comparison of (a) Mechanical and (b) Laser Perforation of Paper [14]

6.6.3 Glass

Conventional methods for glass cutting consist of scratching and breaking or grinding. Complex shapes are difficult to produce using this technique and the cutting speed is generally low. To produce complicated shapes quickly, drilling and cutting of glass plates can be performed with a CO_2 laser synchronized with workpiece movement. The main problem associated with laser glass cutting is thermal cracking or destruction of the workpiece due to the high heat capacitance and low thermal conductivity of the material. In some instances the bending strength of laser-cut glass samples may be reduced to only 50% to 67% that of mechanically-cut glass due to microcracking. There are two methods for laser cutting of glass: controlled fracturing and melting [2]. Laser controlled fracture can be accomplished with low power (on the order of 100W), but the kerf quality depends strongly on the state of stress of the workpiece prior to laser interaction. Residual stresses in the workpiece may cause secondary fracturing away from the kerf when the workpiece is heated by the beam; this effect reduces the straightness of the kerf edge. Laser cutting through melting usually requires laser power above 1kW, but produces smooth and clean kerf surfaces due to the resolidified molten glass. Since material removal occurs through melting, a high-pressure gas jet is found to be effective for ejecting molten glass from the kerf [59]. For a laser power of 1.2kW, glass plates up to 5mm in thickness for both regular and crystallized glass can be cut [59]. Thin-walled glass tubing can be cut using a lower laser power (250W) [14]. The resulting tube sections have smooth and polished ends. Laser cutting also produces minimal residual stresses. In general, good surface quality can be achieved by using high beam power and high scanning velocity. For cutting of intricate shapes a low scanning velocity is required; in this case a pulsed beam can be used to minimize microcracking. Microcracks can also be minimized by preheating the workpiece to reduce laser-induced thermal stresses.

Laser scribing of glass can be performed [26] using a pulsed beam in order to initiate a fault for separation of the glass through fracture by bending. This scribing technique is similar to chemical or diamond etching. By using a beam with short pulses (30ms duration per pulse), a groove composed of a number of shallow cavities 0.08mm to 0.1mm deep can be formed. Microcrack formation can be observed at the groove surface. In some applications, the microcracks are a desirable characteristc due to their refraction of light and highlighting of the kerf area.

6.6.4 Rubber

Due to the high elasticity of rubber materials, the cutting force generated by a mechanical machining operation may cause significant part distortion and dimensional inaccuracies. Mechanical cutting with a blade causes elastic deformations, so the final part geometry may not be equal to the desired geometry. Deformations may be large for high-porosity materials, such as foam rubber. Laser machining on the other hand produces no cutting forces on the workpiece, so dimensional accuracy and repeatability are improved. Also, most rubber materials have high absorptivity at the CO_2 laser wavelength, so energy efficiency is high.

Laser drilling of silicone rubber was investigated in [14] using both stationary beam and trepanning methods. With a stationary beam, 0.43mm diameter holes were drilled in 0.3mm thick silicone rubber with a cycle time of 20ms per hole using a 185W beam. With a trepanning method, a 375W beam can be used to drill 4mm diameter holes in the same material with a cycle time of 100ms per hole. 1.5mm thick foam rubber can also drilled with a 375W beam by trepanning, with hole diameters between 1.5mm to 12mm and cycle times between 20ms to 200ms per hole.

High-speed laser drilling can also be performed by scanning a continuous beam over a perforated metallic mask. A 375W beam was used in conjunction with a brass mask to produce 0.2mm diameter holes in 2mm thick silicone rubber at a rate of 20ms per hole [14].

Typical applications of laser machining for rubber include laser drilling of holes for flowmeter valves, bottle nipples, and catheters. A pulsed CO_2 laser can pierce bottle nipples at a rate of up to 250 parts per minute [18]. Laser scribing is also used to engrave rubber printing plates. An optical reader can be used to scan the image to be reproduced, and a CO_2 laser can be used to engrave the image on a rubber film. This process results in a seamless print roller.

Material	Laser Type	Laser Power	Cutting or Drilling Speed	Kerf Depth	Kerf Width	Application	Refs
Wood (Soft Maple)	CO_2	250W	1m/min	13mm	0.6mm		38
Wood (Hard Maple)	CO_2	250W	0.28m/min	22mm	0.6mm		38
Wood (Red Oak)	CO_2	250W	0.06m/min	29mm	0.6mm		38
Plywood	CO_2	400W		19mm		Dieboard Cutting	38
Paper	CO_2	185W	330m/min	0.2mm		Cigarette Paper Cutting	14
Glass	CO_2	1.2kW		5mm		Plate Glass Cutting	59
Glass	CO_2	250W		1mm		Glass Tube Cutting	59
Silicone Rubber	CO_2	185-375W	20-100ms per hole	0.3mm	0.4-4mm		14

TABLE 6.14   Representative Data for Laser Machining of Wood, Paper, Glass, and Rubber

6.7   Micromachining

With the development of miniaturization of electronic components in recent years, an emerging field for laser machining is the area of micromachining.   Typical micromachining processes produce kerfs with depth and width less than 100μm Although CO_2 and Nd:YAG lasers are both used in micromachining, pulsed Nd:YAG lasers are more common due to their high energy densities and small focussed spot (to 2μm diameter in some applications) [5].   The physical mechanisms which occur during micromachining are [18]:

- Beam energy absorption by the film
- Beam energy reflection and absorption by the substrate
- Conduction from film to substrate
- Conduction from substrate to film
- Plasma coupling and explosive effects

One application for micromachining is resistor trimming in thick and thin-film circuits. These circuits are usually produced through a screen printing process, where a ceramic-metallic hybrid ink is squeezed through a screen pattern [36]. Disturbances during the printing process, such as changes in screen thickness, squeeze pressure and ink composition, cause variations in resistor size and correspondingly, resistance values of up to ±15%. One procedure to minimize this variation is to print all conductor paths larger than nominal, then "trim" the paths (i.e. reduce the size) until the correct resistance value is measured. Typically, a Nd:YAG laser is used in a pulsed mode to cut channels into the resistor surface through vaporization. The laser has to be combined with a high-precision workpiece positioning system (Fig. 6.50). The resistance value is measured and compared with the desired value between beam pulses [37] and several resistor trimming patterns can be used to achieve the desired resistance value (Fig. 6.51), whereas the resistance value varies as a function of trim length (Fig. 6.52). Kerf widths usually range between 10μm and 100μm. The physical mechanisms of this process are similar to those of laser scribing. A thin molten layer is often formed during trimming, similar to scribing of metals and ceramics. With this method, resistance variations within ±0.01% of nominal value can be achieved at trimming rates up to 100 resistors per hour.

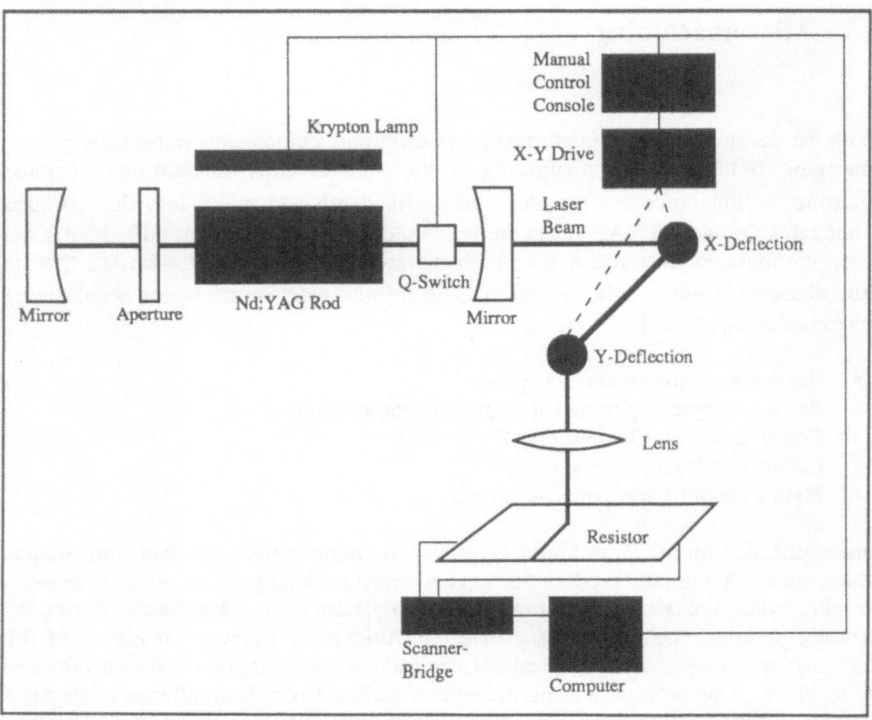

FIGURE 6.50   Laser Configuration for Resistor Trimming [36]

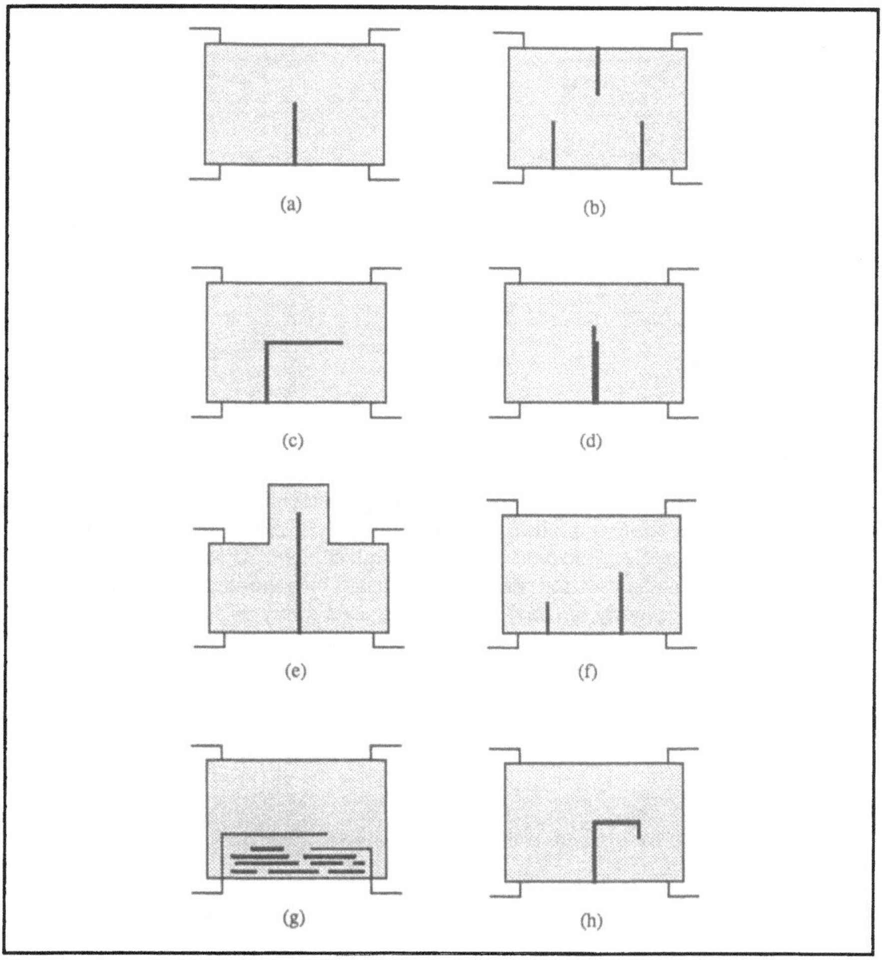

FIGURE 6.51 Resistor Trimming Geometries [36]

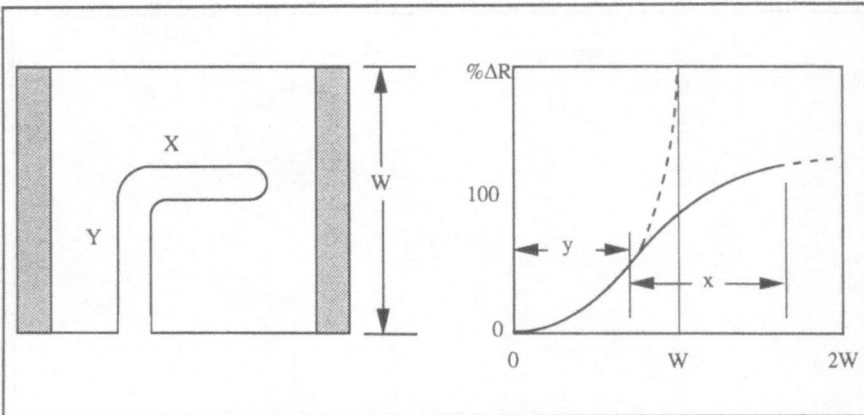

FIGURE 6.52 Change in Resistance vs. Trim Length [36]

Another application of laser micromachining is in the labeling of silicon wafers. This process is identical to laser marking processes used for metal or ceramic parts. An identification number is scribed onto each wafer with either a CO_2 or Nd:YAG laser [37]. A laser beam can also be used to repair metallic masks by vaporizing portions to alter the pattern (Fig. 6.53). Finally, a laser beam can be used to repair a circuit by vaporizing defective components and redirecting channels to circumvent the defective component.

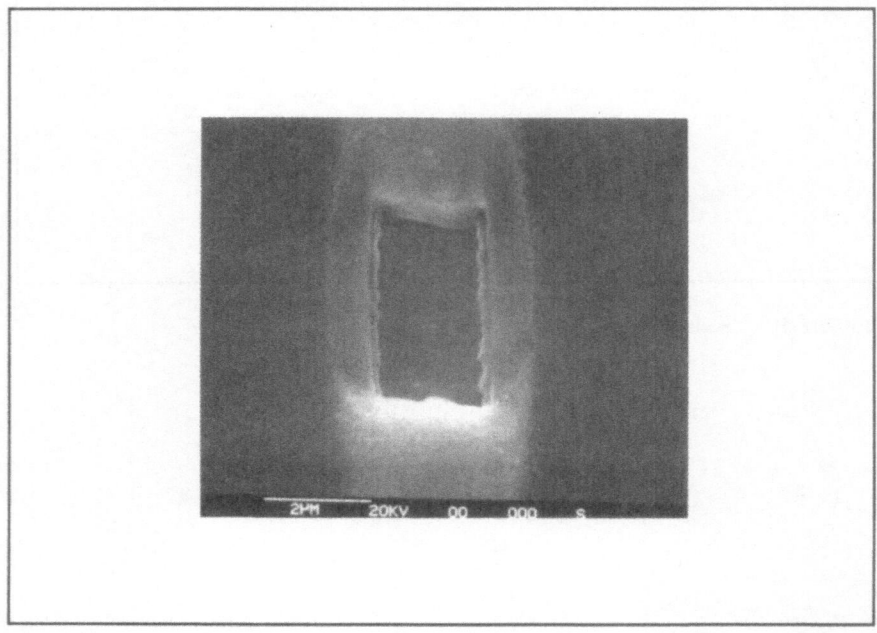

FIGURE 6.53 Laser-Machined Hole in Conductor Path [37]

Laser drilling can also be performed on ferrite materials (such as MnZn and NiZn ferrites) to shape magnetic recording heads for disk and tape drives [45]. Here, the thermal stresses induced by laser drilling are of particular interest, since the stress field may alter the magnetic characteristics and performance of the finished parts. A Nd:YAG laser was used with 2W beam power and a focussed diameter of 20μm [45]. Pulse frequencies range between 1kHz and 4 kHz, and pulse duration was 2ns. A low scanning speed was used (0.1 to 1 mm/s). After drilling, approximately 60-80% of the hole volume was filled with resolidified material, which increases the surface roughness of the part. The resolidified material also represents regions of high stress concentration due to material shrinkage during cooling and may lead to microcrack formation.

Material	Laser Type	Laser Power	Scanning Speed	Groove Depth	Groove Width	Application	Refs
Cermet	Nd:YAG	10^9W/cm^2		15μm	3μm	Thick Film Resistor Trimming	5.36
Cermet	CO$_2$	2W		15μm	30μm	Thick Film Resistor Trimming	5.36
EMCA 5000	Nd:YAG	100kW/cm^3	21.6mm/s	10μm - 17μm	0.38mm	Resistor Trimming	5.12
MnZn Ferrite	Nd:YAG	2W	0.1mm/s - 1mm/s	15μm		Magnetic Head Shaping	5.45
NiZn Ferrite	Nd:YAG	2W	0.1mm/s - 1mm/s	15μm		Magnetic Head Shaping	5.45

TABLE 6.15   Representative Data for Laser Micromachining

References

1.   Affolter, P., and H.G. Schmid, "Processing of New Ceramic Materials With Solid State Laser Radiation," *SPIE – High Power Lasers and Their Industrial Applications,* Vol. 801(1987), 120-129.

2.   Anatasov, P.A. and S.I. Gendjov, "Laser Cutting of Glass Tubing - A Theoretical Model," *J. Physics D: Applied Physics* (1987), 597-601.

3.   Arata, Y., H. Maruo, I. Miyamoto, and S. Takeuchi, "Improvement of Cut Quality in Laser-Gas-Cutting Stainless Steel," *Proceedings of the First International Laser Processing Conference* (1981).

4.   Babenko, V.P. and V.P. Tychinskii, "Gas-Jet Laser Cutting (Review)," *Soviet Journal of Quantum Electronics,* Vol. 2, No. 5 (Mar.-Apr. 1973), 399-410.

5.   Bube, K.R., A.Z. Miller, A. Howe, and B. Antoni, "Influence of Laser-Trim Configuration on Stability of Small Thick-Film Resistors," *Lasers in Modern Industry* , Soc. of Mfg. Eng., Dearborn, MI, 1979, 245-250.

6.   Chryssolouris, G. and J. Bredt, "Machining of Ceramics Using a Laser Lathe," *International Ceramics Review* (April 1988), 70-72.

7.   Chryssolouris, G., J. Bredt, S. Kordas and E. Wilson, "Theoretical Aspects of a Laser Machine Tool," *ASME Journal of Engineering for Industry,* Vol. 110 (Feb. 1988), 65-70.

8.   Chryssolouris, G. and W. Choi, "Gas Jet Effects on Laser Cutting," *Proceedings, SPIE Conference on CO_2 Lasers and Applications* (1989).

9.   Chryssolouris, G. and W. Choi, "Theoretical Aspects of Laser Grooving," *Fourteenth Conference on Production Research and Technology* (1987), 323-331.

10. Chryssolouris, G. and S. Kyi, "Laser Fracture Cutting of Ceramics," Laboratory for Manufacturing and Productivity working paper.

11. Chryssolouris, G., P. Sheng, and W.C. Choi, "Analysis of the Laser Grooving Process for Ceramic and Composite Materials," *Proceedings, Fifteenth Conference on Production Research and Technology* (1988).

12. Chryssolouris, G., P. Sheng, and W.C. Choi, "Investigation of Laser Grooving for Composite Materials," *Annals of the CIRP* (1988), 161-164.

13. Chryssolouris, G., P. Sheng, and W.C. Choi, "Three-Dimensional Laser Machining of Composite Materials," *PED-35 Machining Composites,* Amer. Soc. of Mech. Eng., 1988, 19-30.

14. Coherent, Inc., *Lasers–Operation, Equipment, Application, and Design,* McGraw-Hill, New York, 1980, 137-196.

15. Copley, S., "Laser Machining Ceramics," *Proceedings of the First International Laser Processing Conference* (1981).

16. Copley, S., M. Bass, B. Jau, and R. Wallace, "Shaping Materials With Lasers," *Laser Material Processing* (1983), 297-336.

17. Corfe, A.G., "Laser Drilling of Aero Engine Components," *Proceedings, First International Conference on Lasers in Manufacturing* (1983), 31-40.

18. Duley, W.W., *Laser Processing and Analysis of Materials* , Plenum Press, New York, 1983.

19. Eberhardt, G., "Survey of High Power CO_2 Industrial Laser Applications and Latest Laser Developments," *Proceedings, First International Conference on Lasers in Manufacturing* (1983), 13-19.

20. Engel, S., "Laser Cutting of Thin Materials," *Lasers in Modern Industry,* Soc. of Mfg. Eng., Dearborn, MI, 1979, 207-216.

21. Firestone, R.F. and E.J. Vesely Jr., "High Power Laser Beam Machining of Structural Ceramics," *ASME Symposium on Machining of Advanced Ceramic Materials* (1988), 215-227.

22. Flaum, M. and T. Karlsson, "Cutting of Fiber-Reinforced Polymers With CW CO_2 Laser," *SPIE–High Power Lasers and Their Industrial Applications,* Vol. 801 (1987), 142-149.

23. Garman, G. and J. Ponce, "YAG Laser Marking," *Industrial Laser Annual Handbook,* Penwell Pub., Tulsa, OK, 1986, 121-131.

24. Hamann, C. and H. Rosen, "Laser Machining of Ceramic and Silicon," *SPIE– High Power Lasers and Their Industrial Applications,* Vol. 801 (1987), 130-137.

25. Hamilton, D.C., "CO_2 Laser Marking," *Industrial Laser Annual Handbook,* Penwell Pb., Tulsa, OK, 1986, 132-138.

26. Harrysson, R. and H. Herbertsson, "Machining of High Performance Ceramics and Thermal Etching of Glass by Laser," *Proceedings, Fourth International Conference on Lasers in Manufacturing* (1987), 211-220.

27.   Joeckle, R. and A. Sontag, "Glass Working With CO2 and HF Lasers," *SPIE - High Power Lasers and Their Industrial Applications,* Vol. 801 (1987), 138-141.

28.   Konig, W., C. Wulf, P. Grab, and H. Willerscheid, "Machining of Fibre Reinforced Plastics," *Annals of the CIRP* (1985), N2.

29.   Laudel, A., "Laser Machining of Ceramic," DOE Rept. No. BDX-613-2507 (1980).

30.   Lee, C.S., A. Goel and H. Osada, "Parametric Studies of Pulsed Laser Cutting of Thin Metal Plates," *Journal of Applied Physics,* Vol. 58, No. 3 (Aug. 1985), 1339-1343.

31.   LIA Laser-Material Processing Committee, "Laser Drilling," *Lasers in Modern Industry*, Soc. of Mfg. Eng., Dearborn, MI, 1979, 203-206.

32.   Luxon, J.T. and D.E. Parker, *Industrial Lasers and Their Applications* (1985), Prentice-Hall, Engelwood Cliffs, NJ, 200-242.

33.   Mello, M.D., "Laser Cutting of Non-Metallic Composites," *Proc. SPIE - Laser Processing: Fundamentals, Applications, and Systems Engineering* (1986), 288-290.

34.   Nielson, S.E., *Laser Cutting with High Pressure Cutting Gases and Mixed Gases*, PhD Thesis, Technical University of Denmark (1985).

35.   Nilsson, K. and I. Sarady, "Cutting and Welding Three-Dimensional Sheet Metla Parts with High Accuracy," *Laser 83 Opto-Elektronik.*

36.   Oakes, M., "An Introduction to Thick Film Resistor Trimming by Laser," *Lasers in Modern Industry*, Soc. of Mfg. Eng., Dearborn, MI, 1979, 237-244.

37.   Parker, D.L., "Laser Production Applications in Microelectronics," *Industrial Laser Annual Handbook* , Penwell Pub, Tulsa, OK, 1986, 139-146.

38.   Peters, C.C., "Cutting Wood Materials by Laser," U.S.D.A. Forest Service Paper FPL250 (1975).

39.   Petring, D., P. Abels, E. Beyer, and G. Herziger, "Werkstoffbearbeitung mit Laserstrahlung," *Feinwerktechnik,* Vol. 96 (1988), 264-272.

40.   Polk, D.H., C.M. Banas, R.W. Frye, and R.A. Gragosz, "Laser Processing of Materials," *Industrial Heat Exchangers* (1986), 357-364.

41. Powell, J., G. Ellis, I.A. Menzies, and P.F. Scheyvaerts, "CO_2 Laser Cutting of Non-Metallic Materials," *Proceedings, Fourth International Conference on Lasers in Manufacturing* (1987), 69-82.

42. Schulz, W., G. Simon, H.M. Urbassek, and I. Decker, "On Laser Fusion Cutting of Metals," *Journal Physics D: Applied Physics* ,Vol. 20 (1987), 481-488.

43. Schuocker, D., "Laser Cutting," *Industrial Laser Annual Handbook*, Penwell Pub., Tulsa, OK, 1986, 87-107.

44. Sepold, G. and R. Rothe, "Laser Beam Cutting of Thick Steel," *ICALEO*, SPIE, 1983, 156-159.

45. Siekman, J.G. "Analysis of Laser Drilling and Cutting Results on Al_2O_3 and Ferrites," *Journal of the American Ceramics Society* (1979), 225-231.

46. Steen, W.M. and J.N. Kamulu, *Materials Processing: Theory and Practice Vol. 3*, IFS, Berlin, Germany, 1983, 18-29.

47. Sturmer, E. and M. von Allmen, "Influence of Laser-Supported Detonation Waves on Metal Drilling With Pulsed CO2 Lasers," *Journal of Applied Physics*, Vol. 49, No. 11 (Nov. 1978), 5648-5654.

48. Tagliaferri, V., I. Crivelli Visconti, and A. Di Ilio, "Machining of Fibre Reinforced Materials With Laser Beam: Cut Quality Evaluation," *Proceedings, Sixth International Conference on Composite Materials* (1987), 1.190-1.198.

49. Tagliaferri, V., A. Di Ilio, and I. Crivelli Visconti, "Laser Cutting of Fibre-Reinforced Polyesters," *Composites*, Vol. 16, No. 4 (Oct. 1985), 317-326.

50. Treusch, H.G. and G. Herziger, "Metal Precision Drilling With Lasers," *SPIE - High Power Lasers and Their Industrial Applications*, Vol. 650 (1986), 220-225.

51. Van Cleave, R.A., "Characteristics of Laser Cutting Kevlar Laminates," DOE Rept. No. BDX-613-2075 (1979).

52. Van Cleave, R.A., "Laser Cutting Plastics," DOE Rept. No. BDX-613-2906 (1983).

53. Van Cleave, R.A., "Laser Cutting Plastic Materials," DOE Rept. No. BDX-613-2476 (1980).

54. Van Cleave, R.A., "Laser Cutting Shapes in Plastics," DOE Rept. No. BDX-613-2727 (1981).

55.   Van Dijk, M., "Pulsed Nd:YAG Laser Cutting," *Industrial Laser Annual Handbook*, Penwell Pub., Tulsa, OK, 1987, 52-64.

56.   Wagner, R.E., "Laser Drilling Mechanics," *Journal of Applied Physics,* Vol. 45, No. 10 (Oct. 1974), 4631-4637.

57.   Wallace, R., M. Bass, and S. Copley, "Curvature of Laser-Machined Grooves in Si_3N_4," *Journal of Applied Physics,* Vol. 59, No. 10 (May 1986), 3555-3560.

58.   Willis, J.B., "Techniques and Applications of Laser Marking," *Proceedings, First International Conference on Lasers in Manufacturing* (1983), 53-62.

59.   Yamazaki, K. and H. Kanenatsu, "A Study on Laser Cutting of Glass Plate," *Proceedings, International Conference on Laser Advanced Materials Processing* (1987), 208-215.

60.   Yilbas, B.S., "Investigation into Drilling Speed During Laser Drilling of Metals," *Optics and Laser Technology*, Vol. 20, No. 1 (Feb. 1988), 29-32.

61.   Yilbas, B.S., "Study of Affecting Parameters in Laser Hole Drilling of Sheet Metals," *ASME Journal of Engineering Materials and Technology*, Vol. 109 (Oct. 1987), 282-285.